Everyday Mathematics®

The University of Chicago School Mathematics Project

Minute Math® +

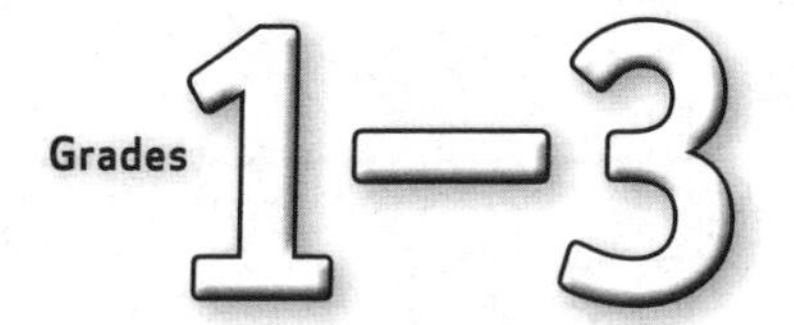

Chicago, IL • Columbus, OH • New York, NY

everyday**math**.com

Send all inquiries to:
McGraw-Hill Education
P.O. Box 812960
Chicago, IL 60681

ISBN: 978-0-07-657722-4
MHID: 0-07-657722-8

Printed in Mexico

4 5 6 7 8 9 DRN 17 16 15 14 13 12

Developed by
Max Bell
Jean Bell
Sheila Sconiers

Contributors
Mary Fullmer
Rosalie Fruchter
Curtis Lieneck

Photo Credits
Cover (l)Ralph A. Clevenger/CORBIS, (c)Estelle Klawitter/CORBIS, (tr)Tim Flach/Stone+/Getty Images.

CONTENTS

Introduction i
Basic Routines 1
Minute Math Topics 25
- Counting 27
- Operations 39
- Geometry 53
- Measurement 61

Number Stories 77
List of Activities by Page 161
Key to Sources 169
Bibliography 173

Introduction

Minute Math®+ is a collection of short mathematics activities that require little preparation. Most of the activities do not require pencil and paper, chalkboard, measuring tools, or manipulative materials. Hence, these activities can be done anywhere. Most of them are brief enough to do in a minute.

The *Minute Math*+ activities are part of the University of Chicago School Mathematics Project's *Everyday Mathematics*® programs for Grades 1–3 and of the UCSMP MathTools for Teachers staff development program. They can be used in Grades 1–3 as a supplement to any basal mathematics program. They can be used with large or small groups of children at any time of the day. You might, for instance, use them during regular class time, during transitions while waiting for the group to assemble or move from one activity to another, while waiting in lines, at dismissal time, and so on. Teachers are often surprised at how much good mathematics can be learned during these few minutes.

The Role of Minute Math+ in the Classroom

Minute Math+ activities play at least five major roles in Grades 1–3.

1. They provide reinforcement and continuous review of the mathematics content.
2. They provide practice with mental arithmetic and logical thinking activities.
3. They give children additional opportunities to think and talk about mathematics and to try out new ideas on themselves, their teachers, and their classmates.
4. They help promote the process of problem solving, which, in the long run, is more important than getting quick answers. Children become more willing to risk sharing their thoughts and their solution strategies with classmates.
5. They increase the time children spend in learning and reviewing mathematics without increasing the time spent in mathematics lessons. Since they can serve as fillers and transitions at any time during the day, they often put time to good use that would otherwise be wasted.

The Parts of *Minute Math*+

Minute Math+ is divided into six parts. You can easily identify the different parts by the bars and tabs along the outer edges of the pages.

Basic Routines
This part of the book presents a cross-section of sample activities drawn from the parts that follow. It is a good place to begin if you have not previously used *Minute Math* activities. As you gain experience, you can move to selecting your own mix of activities from the main subject-matter topics (Parts 2–5) and from "Number Stories."

Minute Math *Topics*
Four major mathematics topics are presented in Parts 2–5:

Counting

Operations

Geometry

Measurement

Number Stories
This is the longest section of *Minute Math*$^+$, and it can provide for discussions that lead to many good problem-solving activities.

The Organization of the Activities

Each page of *Minute Math*$^+$ begins with a basic activity followed by options for adapting that activity to various age and ability levels. The colored dot patterns in front of these options indicate a progression from easy to more difficult variations. The dots are not intended to indicate grade levels. For example:

> The number is ***5. Double it.*** (10)
>
> ●○○○○ Choose a 1-digit number. Double it.
>
> ●●○○○ Choose a 1-digit number. Triple it.
>
> ●●●○○ Choose a 1-digit number. Quadruple it.
>
> ●●●●○ Choose a 2-digit number. Triple or quadruple it.

The phrases and numbers in italics and on the blanks are offered as examples to help you get started. As indicated by the options, you should make substitutions depending on the age, experience, and abilities of your children.

In the example above, the activity could be stated as "The number is 5.

Double it." or "The number is 8. Triple it." or "The number is 6. Quadruple it." or "The number is 25. Triple it." You can repeat the same activities many times; simply change the numbers and words to provide for the growth of the children.

You will probably find that, in most classes at any age level, there are children who find rewards with success at a basic level and others who appreciate being challenged at a higher level. Most of the activities involve group responses. Hence, at more challenging levels, those who can't respond can benefit without embarrassment from the responses of those who do respond.

Many of the *Minute Math⁺* problems use real data and information about people, animals, nature, and other topics. These numbers are usually interesting and often surprising. A key to sources is provided at the end of the book.

The most important things for you to remember as you use *Minute Math⁺* are that the activities should meet the needs of your children and also be catalysts for your own ideas. Ask the children for their activity ideas. Be adventuresome. Try new things and larger numbers—the children's responses may surprise you.

BASIC ROUTINES

Numbers Before and After

What number comes after (follows) ***8*** when you count?

What number comes before (precedes) ***4*** when you count?

What numbers come before and after ***6***?

●○○○○ Use 1-digit numbers.

●●○○○ Substitute 2-digit numbers, including those that will change the decade. Example: What number comes after 29 when you count?

●●●○○ Ask the child what number comes 2 or 3 (or another number) *before* or *after* a given number. Example: What number comes 2 before 9 when you count?

●●●●○ Substitute 3-digit or 4-digit numbers and ask what number comes 2 or 3 (or another number) *before* or *after* the given number. Examples: What number comes 3 after 391 when you count? What number comes 3 before 391 when you count?

Numbers Between

Tell me any numbers between ***25*** and ***35.***

●○○○○ Choose whole numbers.

●●○○○ Choose negative numbers. Example: What numbers are between 0 and −4?

●●●○○ Choose fractions or decimals. Example: What are some numbers (fractions) between $\frac{1}{4}$ and $\frac{3}{4}$? (An infinite number of answers are possible.)

Counts and Skip Counts

I will begin counting. If I point to you, continue the number sequence until I stop you and point to someone else. ***15, 20, 25, 30 . . .***

●○○○○ Begin with numbers 1–50. Count by 2, 5, or 10.

●●○○○ Begin with numbers 1–100. Count forward or backward by 1, 2, 5, or 10.

●●●○○ Begin with numbers 100–900. Count forward or backward by 1, 2, 5, 10, 25, or 50.

●●●●○ Begin with negative numbers or with numbers greater than 1,000. Count forward or backward by 1, 2, 5, 10, 25, 50, or 100.

Ordinal Numbers

Who is ***second*** in line? Who is ***tenth*** in line? . . .

●○○○○ Who is first in line? Who is third from the end? Who is third from the beginning? What number is that person from the end? Who is eighth in line?

●●○○○ Identify children by their place in line, and ask them to perform some sort of action. Example: The third child in line and the seventh child in line change places. Is the fourth child in line still the same?

Variation 1 Perform several actions such as jumping, clapping hands, and snapping fingers in sequence. What did I do ***first? Next? Last?*** (or ***first, second, third***). Ask a volunteer to perform another sequence of actions.

Variation 2 Which letter is ***twelfth*** in the alphabet? (L) The letter ***t*** is what letter? (20th) . . .

Count by 10s and 100s

Count with me: ***13, 23, 33, 43 . . .***

●○○○○ Count forward and backward by 10s. Begin with any 2-digit number.

●●○○○ Count forward and backward by 10s or 100s. Begin with any 3-digit number.

●●●○○ Count forward and backward by 10s, 100s, or 1,000s. Begin with any 4-digit number.

●●●●○ Start with any number. Count forward and backward by any power of 10.

What Do I Do?

I have ***4*** but I want **7.** What do I have to do?

Remember to provide a context for the numbers.

●○○○○ Create problems using small whole numbers. Example: I have 6, but I only want 3. What do I have to do? (Subtract, or take away, 3)

●●○○○ Create problems using whole numbers 1–100 divisible by 5 or 10. Example: I have 10, but I want 30. What do I have to do? (Add 20)

●●●○○ Create problems using simple fractions. Example: I have $\frac{1}{4}$, but I want 1 whole. What do I have to do? (Add $\frac{3}{4}$, or get another $\frac{3}{4}$)

●●●●○ Create problems using whole numbers 1–100. Example: I have 42, but I need 64. What do I have to do? (Add 22)

Complements of 10s

What number must you add to ***6*** to get ***10?***

●○○○○ Say a 1-digit number, and ask what must be added to get 10. Example: What must you add to 3 to get 10? (7)

●●○○○ Say a 1- or 2-digit number, and ask what must be added to get to another decade. Examples: What must you add to 26 to get 30? (4) What must you add to 26 to get 50? (24)

●●●○○ Say a 1- or 2-digit number, and ask what must be added to get to the next century. Examples: What must you add to 6 to get 100? (94) What must you add to 26 to get 100? (74)

●●●●○ Say a 3-digit number, and ask what must be added to get to the next 10, 100, or 1,000. Examples: What must you add to 362 to get 370? (8) What must you add to 362 to get 400? (38) What must you add to 362 to get 1,000? (638)

Arithmetic Facts

4 + ***6*** makes how much? What does **7** – ***3*** equal?

Remember to provide a context for the numbers.

●○○○○ Use single-digit addition and subtraction facts. Example: 9 plus 1 makes how much? (10)

●●○○○ Use 2-digit addition and subtraction facts where at least one number is divisible by 10. Example: What is 50 plus 26? (76)

●●●○○ Use single-digit multiplication and division facts. Example: What does 6 times 4 equal? (24)

●●●●○ Use 2-digit addition and subtraction fact extensions. Examples: If 5 + 7 = 12, how much is 5 + 27? (32) How much is 5 + 87? (92) How much is 5 + 687? (692)

Name Collections (Equivalents)

I'm thinking of the number ***10.*** What are some other names for ***10?*** (5 + 5, 11 − 1, 20 ÷ 2, 1 + 9)

●○○○○ Think of a small whole number 1–10.

●●○○○ Think of a larger whole number 10–100.

●●●○○ Think of a fraction.

●●●●○ Think of a negative number.

Variation 1 Think of two (or 3 or 4) numbers which, added together, make 10. (6 + 4, 5 + 5, . . .)

Variation 2 Think of two numbers that have a difference of 10. (12 − 2, 100 − 90, . . .)

Variation 3 Think of names for 10 that involve products, quotients, or fractions. ($\frac{100}{10}$, 10 × 1, $\frac{1}{2}$ of 20, . . .)

More Name Collections (Equivalents)

I'm thinking of ***50¢.*** What are some other names for ***50¢?*** (2 quarters, 5 dimes, 10 nickels, . . .)

●○○○○ Think of a money amount.

●●○○○ Think of a metric measure. Example: What are some other names for $\frac{1}{2}$ meter? (50 cm, 5 dm, . . .)

●●●○○ Think of a fraction. Example: What are some other names for $\frac{1}{2}$? ($\frac{2}{4}$, $\frac{3}{6}$, $\frac{10}{20}$, . . .)

●●●●○ Think of decimal numbers. Example: What are some other names for 0.01? ($\frac{1}{100}$, $\frac{10}{1,000}$, . . .)

●●●●● Think of a large decade or century number. Example: What are some other names for 4,000? (40 hundreds, 400 tens, . . .)

Multistep Problems

Listen carefully to each step of this problem. Raise your hand when you have the answer. ***9 take away 3, plus 4, add 2, minus 1 makes how much?*** (11)

Remember to provide a context for the numbers.

●○○○○ Use addition and subtraction of small whole numbers. Example: Add 5 to 9, subtract 2, add 12, add 6, subtract 15 equals what? (15)

●●○○○ Use addition and subtraction and doubling of 2-digit whole numbers. Numbers divisible by 5 or 10 are easiest. Example: Subtract 10 from 25, add 5, double it, plus 20, minus 10 equals what? (50)

●●●○○ Use addition, subtraction, multiplication, doubling, and halving of small whole numbers. Example: Add 12 and 10, halve it, take away 4, times 2 equals what? (14)

Variation If calculators are available, young children might practice some problems of this sort on the calculators.

How Many 10s, 100s, 1,000s?

How many ***10s*** are in ***100?*** (10)

●○○○○ Tens in hundreds. Example: How many 10s in 200? (20) How many 10s in 800? (80) . . .

●●○○○ Hundreds in thousands. Example: How many 100s in 1,000? (10) in 3,000? (30) . . .

●●●○○ Hundreds in ten-thousands. Example: How many 100s in 10,000? (100) in 25,000? (250) . . .

●●●●○ How many thousands are in 1 million? (1,000)

What's My Rule?

The rule is ***add 5.*** If the input is ***6,*** what is the output? (11)

●○○○○ Make a rule that adds or subtracts a number 1–10.

●●○○○ Make a rule that adds or subtracts a 2-digit number.

●●●○○ Make a rule that multiplies or divides by a number 1–10.

●●●●○ Make a rule that takes a fraction or a percent of another number. Example: The rule is take $\frac{1}{2}$. If the input is 14, what is the output? (7)

Variation 1 Give the input and the output. Ask children to supply the rule. Example: If the input is 8 and the output is 11, what is the rule? (+3)

Variation 2 Give the rule and the output. Children supply the input. Example: If the rule is add 7 and the output is 12, what is the input? (5)

Number Stories

If I choose you, select two numbers and ask someone else to make up an ***addition number story*** using these numbers.

●○○○○ Make up an addition number story.

●●○○○ Make up a subtraction number story.

●●●○○ Make up a multiplication number story.

●●●●○ Make up a number story with an answer that fits a certain characteristic. Example: Make up a number story where the answer is a negative number. Or, make up a number story where the answer is bigger than 15.

Shapes Around Us

What ***2-dimensional*** shapes do you see ***in this room?*** (circles, triangles, squares, rectangles, . . .)

What ***solid*** shapes do you see ***in this room?*** (spheres, pyramids, cylinders, cubes, prisms, cones, . . .)

Also look for shapes in the hall, on the playground, and in other locations. On field trips, have children watch for and "collect" (in their minds) as many of a particular shape as possible. After the trip, children discuss the shapes they found.

Geometry I Spy

I will choose a "spy." That spy will choose an object in the room and tell us what shape it is. For instance, the spy might say, "I spy a ***circle.***" We will ask the spy yes or no questions to try to determine the object.

●○○○○ Prompt children to "spy" 2-dimensional shapes: circles, triangles, squares, rectangles, . . .

●●○○○ Prompt children to "spy" solid shapes: spheres, pyramids, cylinders, cubes, prisms, cones, . . .

Comparing Measurement Units

Is an inch a bigger, a smaller, or a different kind of measure than a foot? (Smaller. An example of "different" would be centimeters versus grams.)

●○○○○ Emphasize units of length; compare centimeters, meters, and kilometers; or inches, feet, yards, and miles.

●●○○○ Emphasize units of weight; compare grams and kilograms, or ounces and pounds.

●●●○○ Emphasize units of volume; compare milliliters and liters; or pints, quarts, and gallons.

●●●●○ Combine units within either the metric or the U.S. customary system before making the comparison. Example: Is 230 centimeters bigger, smaller, or different than 2 meters?

Recent Dates

What is the date of the day before Friday, July 22, 2005? Choose any recent date.

●○○○○ What is the date of the day after? The name of the day before? The name of the day after? The name of the month before? The name of the month after? . . .

●●○○○ What is the date 2 or 3 (or more) days before or after? The name of the day 2 or 3 (or more) days before or after? The name of the month? . . .

●●●○○ What is the date one (or more) weeks before or after? What is the date and name of the day one week and one day before or after? What is the date one (or more) years later? . . .

Unit Conversions

How many ***minutes*** are in an ***hour?***

●○○○○ Choose measures that increase by a single step. Examples: How many minutes are in an hour? How many days are in a week? How many seconds are in a minute?

●●○○○ Choose multiples of measures that increase by a single step. Examples: How many seconds in 2 minutes? How many minutes in 2 hours? How many days in 3 weeks? . . .

●●●○○ Choose measures that increase by more than a single step and their multiples. Examples: How many seconds in an hour? How many days in 2 months? (Discuss the need to specify which months.)

●●●●○ Choose fractional measures as the larger unit. Examples: How many months in half a year? How many minutes in $\frac{1}{4}$ hour? . . .

Variation Substitute measures of distance, volume, or mass for those of time.

How Many Cents?

How many cents do I have if I have ***3 quarters?***

●○○○○ Use 1–10 of the same coin. Example: How many cents do I have if I have 6 nickels? (Count by 5s to reach 30¢.)

●●○○○ Combine 2 coins or use 10 or more of the same coin. Example: How many cents do I have if I have 5 dimes and 2 nickels? (Count by 10s to reach 50, then count on 10 more to reach 60¢.)

●●●○○ Combine 3 coins. Example: How many cents do I have if I have 2 dimes, 3 nickels, and 26 pennies? (61¢; discuss strategies.)

●●●●○ Use bills. Examples: How many cents do I have if I have a ten-dollar bill? (1,000¢) How many cents do I have if I have a five-dollar bill? (500¢)

Place Value

Tell me a ***2-digit*** number with ***8*** in the ***tens place.*** Tell me another number.

●○○○○ Work with 2-digit numbers. Request specific numbers in the tens place or ones place.

●●○○○ Work with 3- or 4-digit numbers. Request specific numbers in the tens, hundreds, or thousands place.

●●●○○ Work with decimals. Request specific numbers in the tenths, hundredths, or thousandths place. Example: Tell me a number with 7 in the hundredths place.

Ten More, Ten Less

What number is 10 more than ***9?*** What number is 10 less than ***23?***

●○○○○ Use 1-digit numbers with 10 more, 10 less.

●●○○○ Use 2-digit numbers with 10 more, 10 less or with 100 more, 100 less.

●●●○○ Use 3-digit numbers with 10, 100, or 1,000 more or less.

●●●●○ Use 4-digit (or more) numbers with 1,000 or 10,000 more or less.

●●●●● Start with negative numbers or let the answer be negative. Examples: What is 10 less than 8? (−2) What is 10 more than −25? (−15)

MINUTE MATH TOPICS

Counting	27
Operations	39
Geometry	53
Measurement	61

Numbers Before and After

I will point to three of you, one after the other. The first child will say a number such as ***16.*** The second says the number after ***16,*** and the third says the number before ***16.*** Then we will do it again.

●○○○○ Use numbers 1–100.

●●○○○ Use numbers 100–1,000.

●●●○○ Use numbers greater than 1,000.

●●●●○ Use negative numbers.

Counting

Whole Numbers Between

What series of whole numbers comes between the numbers ***12*** and ***15?*** Also, reword the question as: What numbers are greater than ***23*** and less than ***28?***

●○○○○ Choose pairs of whole numbers 1–50 within the same decade. Examples: 20 and 25 or 19 and 14.

●●○○○ Choose pairs of whole numbers 1–100 that cross decades. Examples: 64 and 56 or 73 and 68.

●●●○○ Choose pairs of whole numbers 100–5,000 that cross decades or centuries. Examples: 397 and 389 or 397 and 404 . . .

●●●●○ Specify dollars or cents. Example: How many cents are more than 53¢ and less than 64¢?

Variation Name every other number between 30 and 40. (32, 34, 36, 38)

Missing Numbers

Listen carefully and tell me which numbers I am missing. ***40, 35, 25, 20.***

●○○○○ Count by 1s using numbers 1–50. Example: 21, 22, 24, 25.

●●○○○ Count by 2s, 5s, or 10s using a sequence 1–50. Example: 10, 15, 20, 30, 35.

●●●○○ Count backward by 1s, 2s, 5s, or 10s using numbers 1–100. Example: 87, 86, 85, 83.

●●●●○ Count forward or backward by 1s, 2s, 5s, 10s, or 100s using numbers 100–1,000. Example: 332, 334, 336, 338, 342.

How Many?

If there are ***50 ears,*** how many ***people*** are there?

●○○○○ Choose things that come in pairs—ears, hands, feet, arms, . . .

●●○○○ Choose things that come in 10s—fingers, toes, . . .

●●●○○ Choose things that come in 4s—chair legs, table legs, . . .

Continue the Sequence

Raise your hands as I begin to count. I will point to each of you, one at a time. When I point to you, say the next number in the sequence and put down your hand. ***5, 10, 15, . . .***

After each child has a number: Let's line up in number order from largest number to smallest (or smallest number to largest).

●○○○○ Count by 1s, 2s, 5s, or 10s. Begin at any multiple of that number. Example: Begin with the number 35. Count 35, 40, 45; children continue 50, 55, 60, . . .

●●○○○ Count backward by 1s, 2s, 5s, or 10s.

●●●○○ Count forward or backward using numbers such as 3, 4, 6, or 7. Example: Begin 4, 8, 12; children continue 16, 20, . . .

●●●●○ Count forward or backward using sequences of fractions or decimals. Example: Begin $\frac{1}{4}$, $\frac{2}{4}$ ($\frac{1}{2}$), $\frac{3}{4}$, 1, 1 $\frac{1}{4}$; children continue $1\frac{2}{4}$, $1\frac{3}{4}$, . . .

Repeated Digits

Name a number that is written with ***two*** 7s in it. Be sure children say the number correctly.

●○○○○ Use two repeated digits. Examples: 77; 277; 747; 7,736; . . .

●●○○○ Use three repeated digits. Examples: 777; 3,777; 7,277; 867,757; . . .

●●●○○ Use four repeated digits. Examples: 7,777; 477,377; . . .

Variation Repeat the exercise with a number other than 7.

Numbers with *n* Digits

Which is the smallest ***1-digit*** number? Which is the largest? How many ***1-digit*** numbers are there?

●○○○○ Use 1-digit numbers. (0 is smallest; 9 is largest; there are 10 numbers) These are the ones numbers.

●●○○○ Use 2-digit numbers. (10 is smallest; 99 is largest; 90 numbers) These are the tens.

●●●○○ Use 3-digit numbers. (100 is smallest; 999 is largest; 900 numbers) These are the hundreds.

●●●●○ Use 4-digit numbers. (1,000 is smallest; 9,999 is largest; 9,000 numbers) These are the thousands.

●●●●● Use 5-digit numbers. (10,000 is smallest; 99,999 is largest; 90,000 numbers) These are the ten-thousands.

Although numerals such as 01 or 0032 have not been counted here, commend children who consider these possibilities.

Easier Numbers

Change the number ***19*** to the nearest easy number that ends in ***0.***

●○○○○ Use 2-digit numbers and provide children with two choices for their response. Example: Is the number 27 closer to 20 or to 30?

●●○○○ Use 2-digit numbers. Example: Change 27 to the nearest easy number that ends in 0. (Children respond 30.)

●●●○○ Use 3-digit numbers. Example: Change 384 to the nearest easy number that ends in 0. (Children respond 380.)

●●●●○ Use 3- or 4-digit numbers; change to the nearest 100. Example: 364 becomes 400; 4,692 becomes 4,700; . . .

Variation If a chalkboard is available, write numbers on the board. Ask similar questions.

Money and Measure Counts

Let's count nickels from ***15¢: 15¢, 20¢, 25¢, . . .***

●○○○○ Count nickels or dimes forward and back from any amount.

●●○○○ Count quarters or half dollars forward and back from any amount.

●●●○○ Count $\frac{1}{4}$ cups, $\frac{1}{2}$ cups, or $\frac{1}{3}$ cups forward and back from any amount. Example: $\frac{1}{2}$ cup, 1 cup, $1\frac{1}{2}$ cups, 2 cups, . . .

Creating Numbers

Use the digits ***4, 3, and 7.*** (Choose and adjust digits and the number of digits to the problem.)

●○○○○ Create the largest 2-digit number, the smallest 2-digit number, the largest 3-digit number, the smallest 3-digit number, . . .

●●○○○ Create a number with the 3 in the hundreds place, with 3 in the tens place, with 3 in the ones place, . . .

●●●○○ Create any number with 2 decimal places, the largest number with 2 decimal places, the smallest number with 2 decimal places, . . .

●●●●○ Create a fraction that is less than 1, that is greater than 1, that is greater than $\frac{1}{2}$, that is less than $\frac{1}{2}$, . . .

●●●●● Create the largest fraction you can. Create the smallest fraction you can.

Place Value in Metric Measures

How many ***centimeters*** are in ***2 meters?*** (200 cm)

●○○○○ Use centimeters in whole-number meters.

●●○○○ Use centimeters in meters given as decimals. Examples: How many centimeters are in 2.3 meters? (230 cm) How many centimeters are in 0.23 meters? (23 cm)

●●●○○ Use millimeters in whole-number meters, milliliters in whole-number liters, milligrams in whole-number grams, . . . Example: How many milliliters in 3 liters? (3,000 mL)

●●●●○ Use millimeters in decimal meters, milliliters in decimal liters, milligrams in decimal grams, . . . Example: How many milligrams are in 4.56 grams? (4,560 mg)

Variation Reverse the question. Tell children how many of the smaller unit; ask how many of the larger. Example: How many meters in 342 centimeters? (3.42 m)

Guess My Number

I'm thinking of a number greater than ***55*** but less than ***58.*** What might the number be?

or

I'm thinking of a number smaller than ***72*** but greater than ***68.*** What might the number be?

●○○○○ Use a pair of whole numbers 1–100.

●●○○○ Use a pair of whole numbers 100–1,000.

●●●○○ Use a pair of fractions or decimals.

●●●●○ Use a pair of negative integers.

Remember, there are an infinite number of possibilities between any two numbers. Commend children who suggest fractions or decimals.

Missing Parts in Sums and Differences

What do you subtract from ***12*** to get ***4?*** From what number do you subtract ***3*** to get ***6?*** What do you add to ***5*** to get ***8?*** (Vary the format of the questions, and ask children to discuss their strategies.)

Remember to provide a context for the numbers.

●○○○○ Create problems using small whole numbers. Examples: What do you subtract from 9 to get 9? (0) What do you add to 4 to get 7? (3) From what number do you subtract 3 to get 6? (9)

●●○○○ Create problems using whole numbers that are divisible by 5 or 10. Example: What number do you subtract from 50 to get 40? (10)

●●●○○ Create problems using simple fractions. Example: What do you add to $\frac{1}{4}$ to get 1 whole? ($\frac{3}{4}$)

●●●●○ Create problems using whole numbers 10–100.

●●●●● Create problems using money, time, or decimals.

Parts in a Whole

How many ***thirds*** in ***1?*** (3)

●○○○○ Thirds in 1? (3)

●●○○○ Thirds in 2? (6) In 3? (9)

●●●○○ Thirds in 10? (30) In 100? (300)

Variation Substitute any other fraction for thirds.

Fact Families

Here are 3 numbers of a fact family: ***2, 8, and 10.*** What are all the ***addition*** and ***subtraction*** facts of the family? ($2 + 8 = 10$, $8 + 2 = 10$, $10 - 8 = 2$, and $10 - 2 = 8$)

Remember to provide a context for the numbers.

●○○○○ Give children 1-digit numbers from an addition/subtraction fact family.

●●○○○ Give children 1-digit numbers from a multiplication/division fact family. Example: 2, 8, and 16. Children respond $2 \times 8 = 16$, $8 \times 2 = 16$, $16 \div 8 = 2$, and $16 \div 2 = 8$.

●●●○○ Give children numbers from a fraction fact family. Example: $\frac{1}{4}$, $\frac{3}{4}$, and 1. Children respond with the facts $\frac{1}{4} + \frac{3}{4} = 1$, $\frac{3}{4} + \frac{1}{4} = 1$, $1 - \frac{1}{4} = \frac{3}{4}$, and $1 - \frac{3}{4} = \frac{1}{4}$.

Addition and Subtraction Properties of 10s

35 + 10 = ? (45), 50 − 10 = ? (40)

Operations

●○○○○ Add or subtract 10 to or from numbers divisible by 5 or 10. Examples: 30 + 10 = ? (40), 25 − 10 = ? (15) . . .

●●○○○ Add or subtract 10 to or from any 2-digit numbers. Examples: 17 + 10 = ? (27), 63 − 10 = ? (53) . . .

●●●○○ Add or subtract 100 or 1,000 to or from any number. Examples: 23 + 100 = ? (123), 223 + 100 = ? (323)

●●●●○ Add or subtract multiples of 10, 100, or 1,000 to or from any number. Examples: 453 + 40 = ? (493), 5,000 + 898 = ? (5,898)

Multiplication Properties of 10

$3 \times 10 = ?$ (30), $23 \times 100 = ?$ (2,300)

●○○○○ Use multiples of dimes. Example: How much money is 3 dimes? (30¢)

●●○○○ Use multiples of 10. Example: How much is four 10s? (40)

●●●○○ Multiply 1- or 2-digit numbers by 10. Examples: $22 \times 10 = ?$ (220), $34 \times 10 = ?$ (340)

●●●●○ Multiply 3- or 4-digit numbers by 10, 100, or 1,000. Examples: $342 \times 10 = ?$ (3,420), $467 \times 100 = ?$ (46,700)

How Many Tenths, Hundredths, Thousandths?

How many ***tenths in 1?*** (10)

●○○○○ Tenths in 1? (10) In 2? (20) In 3? (30) . . .

●●○○○ Hundredths in 1? (100) In 2? (200) In 20? (2,000) . . .

●●●○○ Thousandths in 1? (1,000) In 2? (2,000) In 30? (30,000) . . .

●●●●○ Ten-thousandths in 1? (10,000) In 2? (20,000) In 42? (420,000) . . .

Digit Arithmetic

Think of a 2-digit number in which ***the sum of the digits is 11.*** (56, 65, 74, 47, 92, 29, . . .)

●○○○○ The sum of the two digits equals a number 1–18. Or, use 3- or 4-digit numbers the sum of whose digits is a number of your choice. Example: What is a 3-digit number the sum of whose digits is 3? (111, 210, 201, 102, 120, 300)

●●○○○ The difference of two digits equals a number 0–9. Example: Think of a 2-digit number in which the difference between the 2 digits equals 5. (16, 61, 27, 72, 38, 83, 49, 94, 50)

●●●○○ The product of the two digits equals a number 0–81. Example: Think of a 2-digit number in which the product of the 2 digits equals 18. (29, 92, 36, 63)

Secret Numbers

Each of you think of a secret number ***1–10.*** When I call on you, ***add 22*** to your secret number, tell us only the answer, and call on someone else to guess your secret number. For instance, if ***Debbie's*** secret number is 4, she will say 26, and then call on someone to guess her secret number.

●○○○○ Choose a 1- or 2-digit number to add to or subtract from the secret number. Example: Add 23 to the secret number.

●●○○○ Choose a 1-digit number to multiply by the secret number. Example: Multiply the secret number by 8.

●●●○○ Choose two numbers and have the student add the first and subtract the second from the secret number. Example: Add 31 to the secret number, then subtract 10.

Variation Provide the whole class with an operation, then have children take turns with a partner thinking of and calculating the secret number.

Siblings

How many people in our class ***have a brother?*** How many ***have a sister?***

●○○○○ Are there more of you with brothers or more with sisters? How many more? There are ***25*** people in this class, yet ***12*** hands were raised for one question or the other. How come? (If less, some don't have siblings. If more, some have both brothers and sisters.)

●●○○○ More of you have ***brothers*** than ***sisters.*** Does that mean that the total number of boy children in your families is more than the total number of girl children? (No, it could happen that all with sisters have many sisters.) How could we find the total numbers of brothers and sisters if we wanted to?

●●●○○ What fraction of you have brothers? Sisters? Ask the boys with sisters: Do all your sisters have a brother? (Some children answer no—forgetting themselves—if there is not another brother.) Ask a similar question of the girls.

Double, Triple, Quadruple

The number is ***5. Double it.*** (10)

●○○○○ Choose a 1-digit number. Double it.

●●○○○ Choose a 1-digit number. Triple it.

●●●○○ Choose a 1-digit number. Quadruple it.

●●●●○ Choose a 2-digit number. Triple or quadruple it.

Parts

What is $\frac{1}{2}$ of ***10?***

●○○○○ Children determine $\frac{1}{2}$ of 1-digit even numbers.

●●○○○ Children determine $\frac{1}{2}$ of 2-digit even numbers.

●●●○○ Children determine $\frac{1}{2}$ of any 1- or 2-digit number.

●●●●○ Children calculate $\frac{1}{3}$ or $\frac{1}{4}$ of 2-digit numbers divisible by 3 or 4. Examples: What is $\frac{1}{3}$ of 27? (9) What is $\frac{1}{4}$ of 16? (4)

●●●●● Children estimate $\frac{1}{3}$, $\frac{1}{4}$, or $\frac{1}{10}$ of 2- or 3-digit numbers. Example: About how much is $\frac{1}{3}$ of 71? (about 23 or 24)

Operations

Estimating Differences

To the nearest ***10,*** about what is the difference between ***34*** and ***82?*** (about 50)

●○○○○ Children estimate the difference to the nearest 10 between two 2-digit numbers.

●●○○○ Children estimate the difference to the nearest 100 between two 3-digit numbers. Example: To the nearest 100, what is the difference between 292 and 586? (about 300)

●●●○○ Children estimate the difference to the nearest 1,000 between two 4-digit numbers. Example: To the nearest 1,000, what is the difference between 2,535 and 8,624? (about 6,000)

Tenths, or 10%

What is $\frac{1}{10}$ of ***100?*** (10)

●○○○○ Ask for $\frac{1}{10}$ of any 2- or 3-digit number divisible by 10.

●●○○○ Ask for either an estimate or an exact answer for $\frac{1}{10}$ of any 2-, 3-, or 4-digit number. Example: What is $\frac{1}{10}$ of 432? (about 43; exactly 43.2)

●●●○○ Ask for $\frac{1}{10}$ of any money amount. Example: What is $\frac{1}{10}$ of \$5.65? (56¢ or 57¢; emphasize that $\frac{1}{2}$¢ does not exist)

Variation What is 10% of 100? (Restate the problems above using 10% instead of $\frac{1}{10}$. Stress the idea that 10% and $\frac{1}{10}$ have the same meaning.)

Imagining Shapes

Imagine a baseball game. What are some shapes you might see?

Variations Imagine a football game, a hockey game, a basketball game, the zoo, the circus, the playground, your bedroom, your kitchen, the grocery store, your lunch, . . . What are some shapes you might see?

Uni-, Bi-, Tri-

What are the differences among a unicycle, a bicycle, and a tricycle?

What are some other words that begin with *uni-*? (unicorn, unison, unite, . . .) That begin with *bi-*? (biathlon, bicycle, biennial, bifocals, . . .) With *tri-*? (triangle, tripod, triathlon, triplet, . . .)

Shapes of Signs

Imagine you are in a car or a bus. What are the shapes of some road signs that you might see?

triangle: yield sign
rectangle: speed limits, warning signs, street names, . . .
octagon: stop sign
circle: railroad crossings

Why do you think road signs are different shapes?

Making Shapes and Angles

●○○○○ Use your fingers, arms, or body to form a ***circle.*** (Also form other shapes such as triangles, rectangles, and so on.)

●●○○○ With a group of 3 or 4 other children, form a circle either standing up or laying down. (Also form polygons such as triangles, rectangles, pentagons, hexagons, septagons, and octagons.)

●●●○○ Using your arms or fingers, form an ***obtuse angle (greater than 90°).*** (Also form right angles and acute angles.)

Variation Ask another group of 3 or 4 children to form a shape surrounding the original shape.

Identifying Line Segments

Look around the ***room*** to identify:

●○○○○ line segments. (Edge of book, the window, the second hand on the clock, a shelf, corners where walls meet or where walls meet the ceiling, . . .)

●●○○○ line segments that are parallel, that are perpendicular, or that intersect.

Describing Shapes

I'm thinking of a ***2-dimensional*** shape that has ***3 sides*** and ***3 angles.*** Does anyone know what the shape is? (a triangle)

●○○○○ Think of a 2-dimensional shape: circle, triangle, square, . . .

●●○○○ Think of a 3-dimensional shape: sphere, pyramid, cylinder, cube, prism, or cone. Example: The shape I am thinking of looks like a tin can. (cylinder)

Walking Shapes

I will choose one child to trace the imaginary outline of a flat shape on the floor while the rest of us guess the shape.

●○○○○ The chosen child can select any shape to trace.

●●○○○ The chosen child will follow the instructions of another child, who will describe how to walk the shape. Example: To describe a triangle, the student might say, "Walk four paces, turn to the right, walk four more paces, turn to the right, and walk back to the starting point."

●●●○○ The chosen child will follow the instructions of another child, who will describe how to walk the shape. The instructions should include directions, degrees, and paces. Example: To describe a square, the student might say, "Begin facing north, walk 3 paces, turn toward the east at a 90° angle, walk 3 more paces, turn right at a 90° angle to the south, walk 3 more paces, turn right at a 90° angle toward the west, and walk 3 more paces."

Class Shapes and Lines

Line up in the shape of a ***square.***

●○○○○ Line up to form a 2-dimensional shape: circle, square, triangle, . . .

●●○○○ Line up to form two parallel lines, two perpendicular lines, two intersecting, nonperpendicular lines, or a right angle.

Measuring Tools

I'm thinking of a measuring tool that can measure the ***amount of time that has passed in seconds, minutes, and hours.*** What is the tool?

●○○○○ Emphasize measures of length using tools such as rulers, tape measures, and meter or yardsticks. Also emphasize measures of time using a clock or calendar and of temperature using a thermometer.

●●○○○ Emphasize measures of weight using tools such as spring scales, bathroom scales, and pan balances.

●●●○○ Emphasize measures of volume using tools such as measuring cups and measuring spoons.

●●●●○ Emphasize measures of area using tools such as rulers, tape measures, and meter or yardsticks.

Variation 1 I'm thinking of a ***meterstick.*** What could I measure with a ***meterstick?*** What units would I use?

Variation 2 I'm thinking of an ***inch.*** What tool could I use to measure ***inches?*** What objects might I want to measure in ***inches?***

Informal Measuring Tools

What could we use to measure the ***length*** of this ***rug*** if we had no ***rulers*** or ***metersticks*** or ***tape measures?*** (A foot, a piece of paper, the length from elbow to fingertip, . . .) What are the advantages and disadvantages of measuring this way?

●○○○○ What could be used to measure the length of different objects?

●●○○○ What could be used to measure time? (The sun, the moon, how hungry we are, . . .) What could be used to measure temperature? (The clothes we choose to wear in order to feel comfortable tell us approximately how warm or cold it is.)

●●●○○ What could be used to measure weight? (Holding things in our hands, putting things in water to see how quickly they sink, . . .)

How Far to Special Dates

Megan's birthday is ***March 16.*** How far away is that?

●○○○○ Choose a special day within the same month. Count days to figure out how far away that day is.

●●○○○ Choose a special day in the same or the following month. How many weeks and how many days away is it?

●●●○○ Choose a special day later in the year. How many days away is it? How many weeks and days away is it? How many months, weeks, and days away is it?

Variation Choose a special day that has passed. How long ago was it?

Making a Dollar

How many ***pennies*** do we need to make $1.00? (100)

●○○○○ How many pennies do we need to make $1.00? Count by 10s to find the answer. (100) How many dimes do we need to make $1.00? (10)

●●○○○ How many quarters do we need to make $1.00? (4 quarters) How many nickels do we need to make $1.00? (20 nickels)

●●●○○ Ask how many more are needed if we have a given amount of the same coin. Example: How many more quarters do we need to make $1.00 if we have one quarter? (3)

●●●●○ Ask how many more are needed if we have a given amount of different coins. Example: How many more nickels do we need to make $1.00 if we have 3 dimes and 5 pennies? (13 nickels)

Money Exchanges

How many ***nickels*** can we get if we have ***31 pennies?*** (6 nickels) (Exchange one type of money for another.)

●○○○○ Exchange pennies for nickels or dimes. Example: How many dimes can we get if we have 46 pennies? (4 dimes)

●●○○○ Exchange one type of small coin for another. Example: How many quarters can we get if we have 6 dimes? (2 quarters)

●●●○○ Exchange one-dollar bills for ten-dollar bills. Example: How many ten-dollar bills can we get if we have $16? (1 ten-dollar bill)

●●●●○ Exchange ten-dollar bills for hundred-dollar bills. Example: How many hundred-dollar bills can we get if we have 14 ten-dollar bills? (1 hundred-dollar bill)

More Money Exchanges

If I trade ***4 dimes*** for ***46 pennies,*** am I getting a good deal? (Yes, since I give 40¢ to receive 46¢.)

●○○○○ Exchange pennies for nickels or dimes. Example: If I trade 9 pennies for 2 nickels, am I getting a good deal? (Yes. I give 9¢, I receive 10¢.)

●●○○○ Exchange one type of small coin for another. Example: If I trade 6 nickels for 4 dimes, am I getting a good deal? (Yes. I give 30¢, I receive 40¢.)

●●●○○ Exchange one-dollar bills for five-dollar or ten-dollar bills. Example: If I trade 14 one-dollar bills for 2 ten-dollar bills, am I getting a good deal? (Yes. I give $14, I receive $20.)

●●●●○ Exchange ten-dollar bills for fifty-dollar or hundred-dollar bills. Example: If I trade 12 ten-dollar bills for 2 hundred-dollar bills, am I getting a good deal? (Yes. I give $120, I receive $200.)

Coin Equivalents

If I have ***24¢,*** what coins might I have? (Possibilities include 2 dimes and 4 pennies, 4 nickels and 4 pennies, . . .)

●○○○○ Use amounts under $1. Example: If I have 47¢, what coins might I have? (Possibilities include: 4 dimes and 7 pennies; 5 nickels, 2 dimes, and 2 pennies; . . .)

●●○○○ Use amounts over $1. Example: If I have $1.13, what might the coins be? (Possibilities include: 4 quarters, 1 dime, and 3 pennies; 11 dimes and 3 pennies; . . .)

●●●○○ Use amounts under $1 and ask for the smallest number of coins possible. Example: What is the smallest number of coins I could use to make 36¢? (3 coins: 1 quarter, 1 dime, and 1 penny)

Money and Time

Would you rather have ***a penny a day for a month***, or ***a nickel a day for a week?*** (30¢ versus 35¢) (Choose different coin amounts per calendar period. Increase the complexity by increasing the multiplication necessary to determine the answer.)

●○○○○ A dime a day for a month or a penny a day for a year? ($3.00 or $3.10 versus $3.65)

●●○○○ A nickel a day for 3 months or a penny a day for a year? ($4.50 versus $3.65)

●●●○○ A quarter a month for 2 years or a dime a day for 2 months? (Both equal about $6.00; if the months are 31 days each, the dimes would equal $6.20.)

How Many?

How many ***dimes*** and ***pennies*** is ***31¢?*** (3 dimes and 1 penny)

●○○○○ How many dimes and pennies for amounts under $1.00? Example: How many dimes and pennies in $0.83? (8 dimes and 3 pennies)

●●○○○ How many dimes and pennies for amounts over $1.00? Example: How many dimes and pennies in $2.47? (24 dimes and 7 pennies)

●●●○○ How many nickels and pennies? Examples: How many nickels and pennies in 56¢? (11 nickels and 1 penny) How many nickels and pennies in $1.13? (22 nickels and 3 pennies)

●●●●○ How many quarters and pennies? Example: How many quarters and pennies is $1.45? (5 quarters and 20 pennies)

Again, How Many?

If I have ***76¢,*** what coins do I have if I have:

1. only dimes and pennies? (7 dimes, 6 pennies)

2. only nickels and pennies? (14 nickels, 6 pennies)

3. quarters, dimes, and pennies? (2 quarters, 2 dimes, and 6 pennies; 3 quarters, 1 penny; . . .)

4. quarters, dimes, nickels, and pennies? (2 quarters, 2 dimes, 1 nickel, 1 penny; 1 quarter, 4 dimes, 2 nickels and 1 penny; . . .)

●○○○○ Choose amounts under $0.50.

●●○○○ Choose amounts under $1.00.

●●●○○ Choose amounts over $1.00.

Telling Time

About what time is it?

●○○○○ Children respond to the nearest hour or half hour.

●●○○○ Children respond to the nearest 5 minutes.

●●●○○ Children respond to the nearest minute.

●●●●○ Children respond to the nearest minute, and then tell how long it is in minutes until the next hour.

Time Zones*

The time in New York is one hour later than the time in Chicago. The time in California is two hours earlier than the time in Chicago. If it is ***3:00 P.M. in Chicago,*** what time is it in ***New York?*** (4:00 P.M.)

●○○○○ Tell children a time in Chicago; ask them for a time in New York or California.

●●○○○ Tell children the time in New York; ask children for the time in California and the time in Chicago. Or, tell children the time in California, and ask what the time is in New York or Chicago.

* Use time comparisons from your time zone.

Understanding Time

It is about ***2:40 P.M.***

●○○○○ What is another name for 2:40 P.M.? (twenty to 3)

●●○○○ What time will it be in half an hour? (3:10 P.M.) What time will it be in 1 hour? (3:40 P.M.) What time was it 1 hour ago? (1:40 P.M.) (Encourage children to use different names for the same time.)

●●●○○ What time was it 12 minutes ago? (2:28 P.M.) What time will it be in 6 minutes? (2:46 P.M.)

Equivalent Measures

After I state an amount, determine the equivalent amount. (Each of the following variations can be made more or less difficult by altering the size of the numbers and how easily they divide into desired unit amounts.)

Variation 1 Money equivalents: 4 nickels equals how many dimes? (2 dimes) 2 dollars equals how many quarters? (8 quarters) . . .

Variation 2 Time equivalents: 1 week is how many days? (7 days) 1 day equals how many hours? (24 hours) 1 month is about how many days? (28, 29, 30, or 31 days) 2 days equals how many hours? (48 hours) 9 minutes equals how many seconds? (540 seconds) . . .

Variation 3 Distance, volume, and weight equivalents: 12 feet is the same as how many yards? (4 yards) 2 pounds is the same as how many ounces? (32 ounces)

Measurement Fractions

What part of an ***hour*** is ***45 minutes?*** ($\frac{45}{60}$ or $\frac{3}{4}$) (Each of the following variations can be made more or less difficult by altering the size of the numbers and how easily they divide into desired unit amounts.)

Variation 1 Measures of time: What part of an hour is 30 minutes?($\frac{30}{60}$ or $\frac{1}{2}$)

Variation 2 Measures of money: What part of a dollar is 50¢? ($\frac{50}{100}$ or $\frac{1}{2}$ of a dollar) 2 dimes? ($\frac{20}{100}$, $\frac{2}{10}$, or $\frac{1}{5}$ of a dollar)

Variation 3 Measures of distance, capacity, or weight: A pound is 16 ounces. What part of a pound is 8 oz? ($\frac{1}{2}$ lb) 4 oz? ($\frac{4}{16}$ or $\frac{1}{4}$ lb)

Measurement Precision

With most of the measuring [illegible] each of the following [illegible] tell the size of the [illegible] measurement.

[illegible] measurement.

V[illegible] [illegible]

V[illegible] [illegible]

NUMBER STORIES

Baby Penguin Meals

Baby penguins eat almost all the time. Penguin parents feed a baby about 2 pounds of food every hour.

●○○○○ About how much does a baby penguin eat in 2 hours? (4 lb) In 5 hours? (10 lb)

●●○○○ A baby penguin has eaten about 10 pounds of food today. About how many hours have passed? (5 hr)

●●●○○ About how much does a baby penguin eat in 1 day (24 hours)? (48 lb) In 2 days? (96 lb)

Sleep Needs

A growing child needs about 10 hours of sleep each night.

●○○○○ If Jorge has been asleep for about 4 hours, how many more hours should he sleep? (6 hr) If Kim has been asleep for about 7 hours, how many more hours should she sleep? (3 hr)

●●○○○ If Alexis goes to bed at 9:00 P.M., at about what time should she get up in the morning? (7:00 A.M.)

●●●○○ If Rachel wants to be up by about 6:00 A.M. to go fishing, at about what time should she go to bed? (8:00 P.M.)

●●●●○ About how many minutes of sleep does a growing child need each night? (600 min)

Thunder

It takes about 5 seconds for the sound of thunder to travel 1 mile.

●○○○○ About how long would it take the sound of thunder to travel 2 miles? (10 sec) About how long would it take the sound of thunder to travel 5 miles? (25 sec) 10 miles? (50 sec)

●●○○○ About how far can the sound of thunder travel in 1 minute? (12 mi) About how far can the sound of thunder travel in $\frac{1}{2}$ minute? (6 mi)

Rug Measures

The rug in Joshua's room is about ***8*** feet wide. It is about ***2*** feet longer than it is wide.

●○○○○ About how long is the rug? (10 ft)

●●○○○ About how many inches wide is the rug? (96 in.) About how many inches long is the rug? (120 in.)

●●●○○ What is the area of the rug? (80 sq ft)

Old Milk

Milk spoils about 1 week after the expiration date on the package.

●○○○○ About how many days after the expiration date does milk spoil? (7 days) If you buy 2 containers of milk, about how long after the expiration date will they spoil? (still about 1 week)

●●○○○ If the date on Mr. Mayer's milk is 2 days ago, in about how many more days will it spoil? (5 days) If the date is 6 days ago, in about how many days will the milk spoil? (1 day)

●●●○○ If today is ***Wednesday*** and Mr. Mayer's milk will spoil in about 3 days, on what day will it begin to spoil? (Saturday)

●●●●○ If the expiration date on the milk is January 26, on what day will the milk probably spoil? (February 2)

Dreams

People dream an average of 5 times a night.

●○○○○ At that average, about how many dreams might you have tonight? (5 dreams) About how many dreams would you have had last night and the night before? (10 dreams)

●●○○○ About how many dreams might you have had since ***Tuesday?*** About how many dreams do you have in 1 week? (35 dreams)

●●●○○ About how many dreams might our whole class have on some Tuesday night? (number of students 3 5)

●●●●○ About how many dreams might you have in 1 month? (150 dreams) About how many dreams might you have in 1 year? (1,800 dreams)

Airplane Pilots

In 1987, the oldest pilot on record was 94-year-old Ed McCarty of Kimberly, Idaho. The youngest pilot was 10-year-old Cody Locke of Mexico.

●○○○○ How many years older than you was Cody Locke? How many years older than you was Ed McCarty?

●●○○○ What was the difference in the pilots' ages? (84 years) If they were both still alive, what would the difference in their ages be? (still 84 years)

●●●○○ In what year was Cody Locke born? (1977 or 1976) In what year was Ed McCarty born? (1893 or 1892)

●●●●○ In what year would Ed McCarty have been 100? (1993) In what year will Cody Locke be 100? (2077)

The Length of a Dollar Bill

A dollar bill is about 6 inches long.

●○○○○ About how long is a ten-dollar bill? (about 6 in. long) Is a dollar bill more or less than 6 inches wide? (less)

●●○○○ About what part, or fraction, of a foot is the length of a dollar bill? (about $\frac{6}{12}$ or $\frac{1}{2}$ foot) About how many dollar bills would you need to put end to end to make 1 foot? (2 dollar bills) About how many dollar bills would you need to put end to end to make 2 feet? (4 dollar bills)

●●●○○ About how many dollar bills would you need to put end to end to make 1 yard? (6 dollar bills) About how many dollar bills would you need to put end to end to make 2 yards? (12 dollar bills)

Koala Bears

At birth, a koala is about 2 centimeters long. By the time it comes out of its mother's pouch, it is about 20 centimeters long.

●○○○○ Show me, with your hands, about how big the koala is when it is born. Show me, with your hands, about how big the koala is when it comes out of its mother's pouch.

●●○○○ How much does a koala grow before it comes out of the pouch? (18 cm)

●●●○○ How many times bigger is the koala when it comes out of its mother's pouch than when it is born? (10 times bigger)

Classroom Counts

How many children are in the classroom today?

●○○○○ How many children are absent? If each student receives a carton of milk for lunch, how many cartons of milk do we need for our class?

●●○○○ If we invite the class next door over for lunch, and they have the same number of students that we have, how many cartons of milk do we need?

●●●○○ If there are _____ classrooms in our school, about how many students are there in our school?

●●●●○ If there are _____ classes for each grade and _____ grades in our school, how many classrooms are in our school? About how many students are in our school? If each student drinks milk at lunch, about how many cartons of milk should be ordered each day?

Hot Dogs and Buns

Hot dogs often come in packages of 10. Buns usually come in packages of 8.

●○○○○ If we had a class picnic, would we need to buy more packages of hot dogs or more packages of buns? (buns) Why? (There are fewer buns per package.)

●●○○○ If we want to buy enough hot dogs and buns for 20 people, how many packages of each do we need to buy? (2 packages hot dogs, 3 packages buns) If we want to buy enough hot dogs and buns for 35 people, how many packages of each do we need to buy? (4 packages hot dogs, 5 packages buns) How many of each would be left over? (5 hot dogs, 5 buns)

●●●○○ What is the least number of packages of hot dogs and buns you would have to buy to have the same number of each? (4 packages hot dogs, 5 packages buns)

●●●●○ If we bought 5 packages of hot dogs, how many packages of buns do we need if we want a bun for each hot dog? (7 packages) How many hot dogs and how many buns would we have? (50 hot dogs, 56 buns) If we bought 8 packages of hot dogs, how many packages of buns do we need? (10 packages) How many of each would we have? (80 hot dogs, 80 buns)

Making Omelets

Most omelets require 3 eggs. For 3-egg omelets:

●○○○○ How many eggs would you need to make 2 omelets? (6 eggs) To make 3 omelets? (9 eggs) . . .

●●○○○ How many omelets can you make from 1 dozen eggs? (4 omelets) From 3 dozen eggs? (12 omelets) . . .

●●●○○ How many dozen eggs would you use in 1 omelet? ($\frac{1}{4}$ dozen) In 2 omelets? ($\frac{1}{2}$ dozen)

Outgrowing Shoes

As a rule of thumb, children aged 6 to 10 outgrow their shoes about every 84 days.

●○○○○ How would this rule of thumb be different for adults? (Although adult feet sometimes get larger, adults do not outgrow shoes nearly as often as children.)

●●○○○ According to this rule, will you need a new pair of shoes about every month? (no) About every 2 months? (no) About every 3 months? (Yes, probably just before the end of 3 months)

●●●○○ According to this rule, about how many times will you outgrow your shoes this year? (4 or 5 times) Do you remember how many times you outgrew your shoes last year?

NUMBER STORIES

Game Time

Samantha and Aisha played a game for about 45 minutes.

●○○○○ For about how long did each of them play? (45 min) Did the girls play for more or less than 1 hour? (less) How much less than an hour? (15 min)

●●○○○ For what fraction, or portion, of an hour did the girls play? ($\frac{45}{60}$ or $\frac{3}{4}$ of an hour)

●●●○○ If they wanted to play the same game twice, about how long would it take them? (90 min or $1\frac{1}{2}$ hr) If they wanted to play the same game 3 times, about how long would it take them? ($2\frac{1}{4}$ hr or 135 min)

Toilet Flushes

Each toilet flush uses about 7 gallons of water.

●○○○○ About how many gallons of water do 2 toilet flushes use? (14 gal)

●●○○○ About how many gallons of water do 3 flushes use? (21 gal) 4 flushes? (28 gal) . . .

●●●○○ About how many quarts of water does each flush use? (28 qt) About how many quarts of water do 2 flushes use? (56 qt) About how many pints of water does each flush use? (56 pt) . . .

●●●●○ How many times would we need to flush a toilet to use about 112 pints of water? (2 times)

●●●●● If there are 5 people in my family and each person flushes the toilet 3 times a day, about how many gallons of water are used each day? (105 gal)

NUMBER STORIES

Packs of Gum

A small pack of gum contains 5 sticks of gum.

●○○○○ If Marcia chews one piece from the pack and gives another piece to David, how many sticks will be left in the pack? (3 sticks) If she gives Karen one of the remaining sticks, how many will be left? (2 sticks)

●●○○○ How many sticks of gum are in 3 packs? (15 sticks) In 5 packs? (25 sticks) . . .

●●●○○ If Greg has 7 sticks of gum, how many packs does he have? (at least 2 packs) If Bob has 16 pieces of gum, how many packs does he have? (at least 4 packs)

Making Apple Juice

As a rule of thumb, you need 3 apples to make 1 glass of apple juice.

●○○○○ About how many apples would you need to make 2 glasses of juice? (6 apples) To make 3 glasses of juice? (9 apples)

●●○○○ About how many apples would we need to make a glass of juice for each person in this classroom?

●●●○○ If we have 1 dozen apples, about how many glasses of juice could we make? (4 glasses) If we have 2 dozen apples, about how many glasses of juice could we make? (8 glasses) . . .

NUMBER STORIES

Left-Handed People

About 1 out of every 6 people is left-handed.

●○○○○ In a group of 6 people, how many people would you expect to be right-handed? (5 people) Would you always find that 5 out of every 6 people are right-handed? (No. An average is not necessarily true in each separate case.)

●●○○○ In a group of 12 people, how many people would you expect to be left-handed? (2 people) In a group of 12 people, how many people would you expect to be right-handed? (10 people) In a group of 20 people, how many people would you expect to be left-handed? (3 or 4 people)

●●●○○ In this class, about how many people would you expect to be left-handed? How many people really are left-handed? Why might the actual number be different from the expected number?

Outdoor Temperatures

It is about ____ outside right now. Use a current thermometer reading. Sometimes use °C, sometimes °F.

●○○○○ If the high temperature yesterday was about 60°F, is today a warmer day or a cooler day than yesterday? About how much warmer or cooler?

●●○○○ If the expected high today is 73°F, about how many more degrees will the temperature have to rise to reach the expected high? If the predicted temperature for tonight is 58°F, about how many degrees will the temperature change from what it is now?

●●●○○ If the temperature in an Arizona desert is usually about 40°F warmer than the temperature here, what is the approximate temperature in that desert?

●●●●○ If the coldest day this winter was about −5°F, about how much warmer is today than the coldest day of the year?

Making Orange Juice

As a rule of thumb, an orange will yield about a half cup of orange juice.

●○○○○ How many oranges would you need to make about 1 cup of orange juice? (2 oranges)

●●○○○ How many oranges would you need to make about 2 cups of orange juice? (4 oranges) To make about 4 cups of orange juice? (8 oranges) About a cup of orange juice for each member of your family?

●●●○○ If we have 10 oranges, about how many cups of orange juice can we make? (5 c) If we have 40 oranges? (20 c) If we have 100 oranges? (50 c)

●●●●○ How many oranges would you need to make about a pint of orange juice? (4 oranges) To make about a quart of orange juice? (8 oranges)

Cooking-Oil Consumption

Suppose that the school cafeteria uses about 6 quarts of cooking oil every day.

●○○○○ About how many quarts of cooking oil does the cafeteria use in 2 days? (12 qt) In 3 days? (18 qt) In 10 days? (60 qt)

●●○○○ About how many quarts of cooking oil does the cafeteria use each week of school? (about 30 qt; the school week has 5 days) About how many quarts of oil does the cafeteria use in a month? (about 135 qt; accept answers from 120 quarts to 150 quarts)

●●●○○ About how many pints of cooking oil does the cafeteria use each week of school? (60 pt) About how many gallons of cooking oil does the cafeteria use each week of school? ($7\frac{1}{2}$ gal)

●●●●○ Would 10 gallons of cooking oil be enough for 2 weeks? (no) Would 15 gallons of cooking oil be enough for 2 weeks? (yes)

Making Potato Salad

According to some cookbooks, you should use about $1\frac{1}{2}$ potatoes and 1 egg per person when you make potato salad.

●○○○○ If you wanted to make enough potato salad for 2 people, how many potatoes should you use? (3 potatoes) How many eggs should you use? (2 eggs)

●●○○○ If you have plenty of eggs, but only 6 potatoes, how much potato salad could you make? (enough for 4 people) If you have 9 potatoes? (enough for 6 people) . . .

●●●○○ About how many potatoes would you need to make enough potato salad for 12 people? (18 potatoes) For 16 people? (24 potatoes) For 25 people? (37.5 potatoes, but most cooks would use either 37 or 38 potatoes. Discuss this.)

A Snail's Pace

The slowest snails in the world move at a speed of about 23 inches per hour.

●○○○○ Do these snails travel more or less than 1 foot in an hour? (more) Do they travel more or less than 2 feet in an hour? (a little less)

●●○○○ About how many inches do these snails travel in 2 hours? (46 in.) In half an hour? (11 or 12 in.)

●●●○○ About how long would it take this kind of snail to get across a table that is 4 feet wide? (about 2 hr) To cross a room that is 12 feet wide? (about 6 hr)

Riddles

Question:	Why is getting up at 3:00 in the morning like a pig's tail?
Answer:	Because it's twirly (too early).
Question:	What did the acorn say when it grew up?
Answer:	Geometry (Gee, I'm a tree).
Question:	What do you call a dead parrot?
Answer:	Polygon.
Question:	Why is "smiles" the longest word in the English language?
Answer:	There's a mile between the two s's.

The Oldest Living Animal

The oldest living animal on record was found in 1982. At that time, this clam was 220 years old. (It was an ocean quahog, *Arctica islandica.*)

●○○○○ If the clam is still alive today, how old is it?

●●○○○ Was the clam alive in 1882? (yes) Was it alive in 1782? (yes) In 1682? (no)

●●●○○ How old was the clam in 1782? (20 years old) In 1772? (10 years old) In 1770? (8 years old) In 1765? (3 years old)

●●●●○ When was the clam born? (probably in 1762) In what year would the clam be 250 years old if it lives that long? (2012)

Lifespan of a Dollar

A U.S. dollar bill has a lifespan of about 18 months.

●○○○○ What do you think people mean when they say "the lifespan of a dollar bill"? (The amount of time the dollar bill is in circulation—from the time it is printed until it is falling apart or removed from circulation by the government) Does the average dollar bill have a lifespan of more or less than 1 year? (more)

●●○○○ Does the average dollar bill have a lifespan of more or less than 2 years? (less) About how many months less than 2 years is the lifespan of a dollar bill? (6 months less) About how many months more than 1 year is the lifespan of a dollar bill? (6 months more)

●●●○○ About how many days does the average lifespan of a U.S. dollar bill last? (about 540 days)

Baby Weights

The average newborn human baby usually doubles its weight in 6 months.

●○○○○ About how much will a baby weigh in 6 months if it weighed about 7 pounds at birth? (14 lb) If it weighed about $7\frac{1}{2}$ pounds at birth? (15 lb) About 8 pounds? (16 lb)

●●○○○ If a 6-month-old baby weighs about 12 pounds, about how much did it weigh at birth? (6 lb) If a 6-month-old baby weighs about 14 pounds, about how much did it weigh at birth? (7 lb) If it weighs about 16 pounds? (8 lb)

●●●○○ If a newborn weighs about 7 pounds, about how much will it weigh when it is 2 years old? (Insufficient data to answer the question; you don't know how much weight an infant will gain after the first 6 months) If a 6-pound newborn did keep doubling its weight every 6 months, about how much would it weigh when it is 3 years old? (almost 200 lb; that is more than the average adult male)

NUMBER STORIES

Planting Flower Bulbs

As a rule of thumb, plant a flower bulb three times as deep as its length.

●○○○○ About how many inches deep should you plant a 1-inch bulb? (3 in.) Show me, with your hands, about how deep this would be. About how many inches deep should you plant a 2-inch bulb? (6 in.)

●●○○○ If I followed this rule and planted a bulb 6 cm deep, how long is the bulb? (2 cm) If I planted the bulb 9 cm deep? (3 cm)

●●●○○ About how many inches deep should you plant a bulb that is $1\frac{1}{2}$ inches long? ($4\frac{1}{2}$ in.) That is $\frac{1}{2}$ inch long? ($1\frac{1}{2}$ in.) . . .

New Books

Every day about 125 new books are published in the United States. About 8 of these books are children's books.

●○○○○ About how many books for adults are published each day? (117 books)

●●○○○ About how many children's books are published in 2 days? (16 children's books) In 3 days? (24 children's books) What is the average number of books published in 2 days? (250 books) In 3 days? (375 books)

●●●○○ About how many children's books are published in 1 week (5 working days)? (40 children's books) In 3 weeks? (120 children's books)

●●●●○ About how many weeks would it take to publish 200 children's books? (5 weeks)

●●●●● On average, what fraction of books published in one day are written for children? ($\frac{8}{125}$)

Patents* for Inventions

Suppose an average of 300 inventors apply for patents every day and about 180 of these inventors are granted patents.

●○○○○ On the average, do more or less than 200 people receive patents in one day? (less) About how many fewer than 200 people receive patents in one day? (20 fewer people)

●●○○○ On the average, do more or less than half of the inventors who take their patents to the U.S. Patent Office receive patents? (more than half)

●●●○○ About how many patents are applied for in 1 week, or 5 working days? (1,500 patents) About how many patents are granted in 1 week? (900 patents) In 2 weeks? (about 3,000 patents applied for; 1,800 granted)

●●●●○ On any given day, what is the ratio of inventors who are granted patents to the inventors who are not granted patents? (180:120 or 18:12 or 3:2) On the average, for every 3 people who do not receive a patent on a given day, how many people do receive patents? (4 or 5 people)

* Briefly discuss what a "patent" is: an assurance that an inventor has the sole right of ownership to his or her invention for a specified period of time.

Elephant Meals

Elephants spend about 16 hours of every day eating.

●○○○○ About how many hours a day do elephants **not** eat? (8 hr) Do elephants spend more of the day eating or not eating? (eating) About how many hours a day do you usually spend eating? How many hours a day do you **not** eat?

●●○○○ About how many hours do elephants spend eating in 2 days? (32 hr) In 3 days? (48 hr) . . .

●●●○○ On average, what fraction, or portion, of the day do elephants spend eating? ($\frac{16}{24}$ or $\frac{2}{3}$) What fraction, or portion, of the day do they spend **not** eating? ($\frac{8}{24}$ or $\frac{1}{3}$)

●●●●○ About how many minutes a day do elephants spend doing things other than eating? (480 min)

Elephant Sleep

Most elephants need only about 2 hours of sleep each day.

●○○○○ About how many hours of sleep do you need each day? About how many more hours of sleep do you need than most elephants?

●●○○○ About how many hours a week does an elephant sleep? (about 14 hr)
About how many hours a month does an elephant sleep? (about 60 hr)

●●●○○ About how many minutes a day does an elephant sleep? (about 120 min)
About how many minutes does an elephant sleep in 2 days? (about 240 min)

●●●●○ About how many hours a year does an elephant sleep? (about 730 hr)
About how many hours a year do you sleep?

Hibernation

The burrow ground squirrel, which lives in Alaska, hibernates for about 9 months of the year.

●○○○○ For about how many months of the year is the burrow ground squirrel active (not hibernating)? (3 months)

●●○○○ For about how many months does this squirrel hibernate in 2 years? (18 months) In 3 years? (27 months) During your lifetime?

●●●○○ For about how many days of the year does the burrow ground squirrel hibernate? (about 270 to 280 days) For about how many days of the year is the burrow ground squirrel not hibernating? (about 90 to 93 days)

●●●●○ For about what portion, or fraction, of the year does the burrow ground squirrel hibernate? ($\frac{9}{12}$ or $\frac{3}{4}$ of the year)

Variation Pose questions about the Siberian chipmunk, which hibernates 7 to 8 months of the year.

NUMBER STORIES

Making Cookies

A cookie recipe calls for 2 cups of flour and 1 cup of sugar for each batch of cookies.

●○○○○ If you wanted to make twice as many cookies, how many cups of sugar should you use? (2 cups) How many cups of flour? (4 cups)

●●○○○ If you wanted to make half as many cookies, how many cups of sugar would you use? ($\frac{1}{2}$ cup) How many cups of flour? (1 cup)

●●●○○ If you have only a half-cup measuring cup, how many half cups of flour do you need for one batch of cookies? (4 half cups) How many half cups of sugar? (2 half cups) If you have only a one-third cup measuring cup, how many one-third cups of flour would you need? (6 one-third cups) How many one-third cups of sugar? (3 one-third cups)

●●●●○ If you want to make twice as many cookies as the recipe makes, but have only a half-cup measuring cup, how many half cups of flour would you need to use? (8 half cups) How many half cups of sugar? (4 half cups)

Types of Bears

Scientists have identified seven different types of bears in the world. Three of these different types can be found in North America.

●○○○○ How many types of bears in the world cannot be found in North America? (4 types)

●●○○○ What portion, or fraction, of the types of bears in the world can be found in North America? ($\frac{3}{7}$) What portion, or fraction, of the types of bears in the world cannot be found in North America? ($\frac{4}{7}$)

The Three-Toed Sloth

The three-toed sloth, also known as the ai (pronounced "eye"), is one of the slowest moving land animals. It moves an average of about 6 to 8 feet per minute, for an average speed of 0.068 miles per hour.

●○○○○ About how many feet does the ai travel in 2 minutes? (12–16 ft)
Can the ai travel 18 feet in 3 minutes? (yes)

●●○○○ About how many yards does the ai travel in 2 minutes? (4–5 yd)
In 5 minutes? (10–13 yd)

●●●○○ Is the ai likely to move 1 mile in 10 hours? (no) in 20 hours? (yes)
If the ai continues to travel at its average speed for 10 hours, about how far is it likely to go? (0.68 mi, or about $\frac{2}{3}$ mi)

Pumping Blood

As a rule of thumb, your heart pushes about 5 tablespoons of blood into your arteries with each beat.

●○○○○ About how many tablespoons of blood are pushed into your arteries in 2 beats? (10 tbs) In 3 beats? (15 tbs) In 10 beats? (50 tbs) . . .

●●○○○ If your heart beats about 60 times a minute, about how many tablespoons of blood does your heart push into your arteries in 1 minute? (300 tbs) In 2 minutes? (600 tbs)

●●●○○ If you were chasing a friend, would your heart push more or less blood into your arteries in one minute than if you were reading a book? (More; when you are active, your heart beats more rapidly than when you are relaxed. Since this means your heart beats faster when you run, more tablespoonfuls of blood would be pushed into your arteries.)

Basketball Player Heights*

In 1989, the tallest basketball player in the National Basketball Association (NBA) was Manute Bol. He measured 7 feet 6 inches in height. The shortest player in the NBA was Tyrone Bogues, who measured a little more than 5 feet 3 inches in height. Both played for the Washington Bullets.

●○○○○ Was the difference in height between these players more or less than 1 foot? (more) Was the difference in height more or less than 2 feet? (more) More or less than 3 feet? (less)

●●○○○ About how many feet and inches over 6 feet tall is Manute Bol? (1 ft 6 in.) About how many inches under 6 feet tall is Tyrone Bogues? (9 in.)

●●●○○ About how much shorter is Tyrone Bogues than Manute Bol? (2 ft 3 in.)

●●●●○ About how many inches tall is Manute Bol? (90 in.) Tyrone Bogues? (63 in.)

* It would be helpful to write the heights on the chalkboard.

Body Muscle

Less than half of the average person's body weight is muscle.

●○○○○ How much of a person's body weight is not muscle? (more than $\frac{1}{2}$)

●●○○○ If a child weighs about 70 pounds, about how much do the child's muscles weigh? (less than 35 lbs)

To be more exact, about $\frac{2}{5}$ of a person's body weight is muscle.

●●●○○ What fraction of a person's body weight is not muscle? (about $\frac{3}{5}$)

●●●●○ For a 100-pound person, about how many pounds are muscle? (40 lbs) About how many pounds are not muscle? (60 lbs)

●●●●● About what percent of a person's weight is muscle? (40%) About what percent is not muscle? (60%)

A Long Monopoly® Game

The longest game of Monopoly® on record lasted about 1,680 hours.

●○○○○ About how long do you think a Monopoly® game would ordinarily last? (probably 1–3 hours) About how many hours longer than an ordinary game was the longest Monopoly® game? (over 1,670 hours)

●●○○○ Did the longest Monopoly® game last more than 1 day? (yes) More than 1 week? (Yes; count by 25 for 7 days. Why is this a good way to estimate the answer?)

●●●○○ About how many days did the longest Monopoly® game last? (70 days) About how many weeks? (10 weeks)

Paper in Trash Heaps

Some people claim that the average 10-ton pile of municipal trash contains 6 to 7 tons of paper.

●○○○○ Is more or less than half of the pile paper? (more) If people begin to recycle more paper, what will happen to the amount of paper in the municipal trash? (It will decrease.)

●●○○○ About how many tons of paper are in a 20-ton pile of trash? (12–14 tons) In a 30-ton pile of trash? (18–21 tons) A 40-ton pile? (24–28 tons)

●●●○○ If there are 20 tons of trash one day and 50 tons of trash a few days later, about how many tons of paper have been added? (18–21 tons of paper) Since there are 2,000 pounds in 1 ton, about how many pounds of paper have been added? (36,000–42,000 lb)

●●●●○ If there are 19 tons of paper in the trash heap, about how much does the whole trash heap weigh? (about 30 tons) Since there are 2,000 pounds in 1 ton, about how much does the trash heap weigh in pounds? (60,000 lb)

●●●●● About what part of a 10-ton pile of trash is paper? (60%–70%, or $\frac{6}{10}$ to $\frac{7}{10}$) About what part of a 20-ton pile of trash is paper? (60%–70%, or $\frac{6}{10}$ to $\frac{7}{10}$; the ratio of paper to total garbage remains constant)

Breathing

People take approximately 12 breaths a minute when they are relaxed.

●○○○○ About how many breaths do people take in 2 minutes if they are relaxed? (24 breaths) In 3 minutes? (36 breaths) In 4 minutes? (48 breaths)

●●○○○ About how many breaths do people take in half an hour if they are relaxed? (360 breaths) In 1 hour? (720 breaths)

●●●○○ During gym class, will people take more or less than 240 breaths in 20 minutes? (More; they would take about 240 breaths in 20 minutes if they were relaxed. When they are active, they take more breaths per minute.)

Variation Ask the children to count how many breaths they take while you time them for one minute. Get an informal consensus for middle value, then ask questions such as those above.

Cricket Volumes

As a rule of thumb, a quart jar will hold about 1,000 crickets.

●○○○○ Would the same jar hold more or less than 1,000 ants? (more than 1,000; ants are smaller than crickets)

●●○○○ About how many quart jars would you need if you wanted to collect 2,000 crickets? (2 jars) To collect 3,000 crickets? (3 jars)

●●●○○ There are 2 pints in a quart and 4 quarts in a gallon. About how many crickets would fit into a pint jar? (about 500 crickets) Into a half-gallon jar? (about 2,000 crickets) Into your lunch milk carton? (about 250 crickets since most school milk containers are half-pint cartons)

NUMBER STORIES

Tall Buildings*

The Willis Tower is about 443 meters tall. The John Hancock Center is about 343 meters tall. The Eiffel Tower is about 300 meters tall.

●○○○○ Which of these buildings is the tallest? (Willis Tower) Which is the shortest? (Eiffel Tower) If we measured each of these buildings in feet, which would have the largest number of feet? (Willis Tower; the tallest building would have the most number of feet regardless of which units are used)

●●○○○ About how much taller than the Eiffel Tower is the Willis Tower? (143 m) About how much taller than the John Hancock Center is the Willis Tower? (100 m) . . . About how much shorter than the John Hancock Center is the Eiffel Tower? (43 m)

●●●○○ About how many centimeters tall is the Willis Tower? (44,300 cm) One decimeter equals 10 cm. How many decimeters tall is the Willis Tower? (4,430 dm)

Variation Heights of some other tall buildings: Great Pyramid of Cheops, 147 m; Empire State Building, 381 m; Aon Building, 343 m.

* It would be helpful to write building heights on the chalkboard.

Walking Rates

The average city dweller walks about 6 feet per second while the average country dweller walks about 3 feet per second.

●○○○○ Who will walk farther in 1 minute, the average city dweller or the average country dweller? (city dweller) How many more feet per second does the average city dweller walk than the average country dweller? (3 ft per sec)

●●○○○ About how many feet does the average city dweller walk in 2 seconds? (12 ft) In 5 seconds? (30 ft) . . . About how many feet does the average country dweller walk in 2 seconds? (6 ft) In 5 seconds? (15 ft)

●●●○○ In one minute, does the average city dweller walk more or less than 200 feet? (more) In one minute, does the average country dweller walk more or less than 200 feet? (less) In 2 minutes, about how far does the average country dweller walk? (360 ft)

●●●●○ In 1 minute, about how many yards does the average city dweller walk? (120 yd) In 1 minute, about how many yards does the average country dweller walk? (60 yd)

Penguin vs. Human Swimmers

Penguins can swim about 15 miles in 1 hour. A fast human can swim about 3 miles in 1 hour.

●○○○○ About how many more miles than a human can a penguin swim in 1 hour? (12 mi)

●●○○○ About how far could a penguin swim in 2 hours? (30 mi) About how far could a human swim in 2 hours? (6 mi; however, it is unlikely that a human could actually swim fast for 2 hours) About how far could a penguin swim in $\frac{1}{2}$ hour? (7–8 mi, or about $7\frac{1}{2}$ mi) About how far could a human swim in $\frac{1}{2}$ hour? (1–2 mi, or $1\frac{1}{2}$ mi)

●●●○○ About how many times faster than a human can a penguin swim? (about 5 times faster)

●●●●○ About how long would it take a penguin to swim 3 miles? ($\frac{1}{5}$ hr, or about 12 min) If a human could continue swimming at a very fast pace, how long would it take the swimmer to swim 15 miles? (at least 5 hr)

Length of a Shrew

The tiny shrew, which looks something like a mouse, has a head and body length of about 1.7 inches (or about $4\frac{1}{4}$ cm), and a tail length of about 1.2 inches (just over 3 cm).

●○○○○ Which is longer, the length of the shrew's head and body or the length of its tail? (the length of its head and body)

●●○○○ About how long is the entire shrew from head to tail? (2.9 in. or $7\frac{1}{4}$ cm) Show me, with your hands, about how long this is.

●●●○○ About how much shorter is the tail of the shrew than the length of its head and body? (0.5 or $\frac{1}{2}$ in. or $1\frac{1}{4}$ cm)

Animal Litters

Hamsters can have litters about once every 4 months. Each litter usually consists of 6–12 baby hamsters.

●○○○○ About how many litters can hamsters have in 8 months? (2 litters) In 13 months? (3 litters) . . .

●●○○○ About how many times a year can a female hamster have a litter? (3 times) About how many times in 2 years can a female hamster have a litter? (6 times) . . .

●●●○○ About how many babies can a hamster have in 1 year? (18–36 babies)

Variation Pose questions about the birth rates of other animals: red squirrels have 1–3 litters a year with 3–7 young per litter; wood mice have 3–4 litters a year with 4–6 young per litter; foxes average 1 litter a year with 3–8 young per litter.

Washing and Drying Clothes

It takes my washer about 30 minutes to wash clothes, and my dryer about 45 minutes to dry clothes.

●○○○○ About how much longer does it take me to dry clothes than to wash them? (15 min)

●●○○○ About how long will it take me to wash and dry a load of clothes? (75 min, or 1 hr 15 min) To wash and dry 2 loads of clothes? (150 min, or 2 hr 30 min)

●●●○○ My neighbor's washing machine washes clothes about 5 minutes faster than mine, but he has the same dryer as I do. How long will it take him to wash and dry a load of clothes? (70 min, or 1 hr 10 min)

●●●●○ I have about 2 hours of free time. How many loads of laundry can I wash and completely dry? (only 1 load of laundry; I would not have time to dry the second load) If I have about 3 hours of free time? (2 loads) . . .

Growing Eyelashes

On the average, a person grows a new eyelash every 3 months.

●○○○○ About how many months would it take to grow 2 new eyelashes? (6 months) To grow 3 new eyelashes? (9 months)

●●○○○ On the average, about how many new eyelashes does a person grow in one year? (4 new eyelashes) In 2 years? (8 new eyelashes)

●●●○○ On the average, about how many weeks does it take to grow a new eyelash? (about 12 weeks) To grow 4 new eyelashes? (about 48 weeks)

●●●●○ On the average, about how many days does it take to grow 10 new eyelashes? (about 900 days)

Animal Weights*

The average rat weighs about 1 pound. The average chicken weighs about 7 pounds. The average cat weighs about 14 pounds.

●○○○○ About how much heavier is a chicken than a rat? (6 lb) About how much lighter is a chicken than a cat? (7 lb) About how much lighter is a rat than a cat? (13 lb)

●●○○○ About how much would 2 chickens weigh together? (14 lb) About how much would 1 cat and 1 rat weigh together? (15 lb) . . .

●●●○○ If we put 1 chicken on one side of a balance, how many rats would we need on the other side to make it balance? (7 rats) If we put 2 chickens on one side of a balance, how many rats would we need on the other side to make it balance? (14 rats) If we put 1 cat on one side of a balance, how many chickens would we need on the other side to make it balance? (2 chickens)

●●●●○ Which would weigh more, 15 rats or 2 chickens? (15 rats) Which would weigh less, 2 cats or 2 chickens and 5 rats? (2 chickens and 5 rats)

Variation Pose questions using other typical animal weights: guinea pig, 1.5 lb; ferret, 2 lb; fox, 14 lb; sheep, 150 lb; moose, 800 lb; coyote, 75 lb.

* It would be helpful to write the animal weights on the chalkboard.

Body Water Weight

About half a person's weight is water.

●○○○○ About how much of a person's weight is **not** water? (half)

●●○○○ Kim weighs 60 pounds. About how much does the water in his body weigh? (30 lb) Lisa weighs 50 pounds. About how much does the water in her body weigh? (25 lb)

●●●○○ About what percent of a person's weight is water? (50%) About what percent of a person's weight is **not** water? (50%)

Milk Consumption

Imagine that the nursery school down the street uses about 5 quarts of milk every day.

●○○○○ About how many quarts of milk do they use in 2 days? (10 qt) In 4 days? (20 qt) . . .

●●○○○ About how many quarts of milk do they use in 5 days? (25 qt) In 1 week? (25 qt, since they only use milk on the weekdays) In 10 days? (50 qt) . . .

●●●○○ About how many pints of milk do they use each day? (10 pt) In 3 days? (30 pt) . . .

●●●●○ If each child drinks about a half pint of milk each day, how many children are in the nursery school? (20 children)

NUMBER STORIES

Herding Sheep

A sheep farmer needs 1 sheepdog for every 100 sheep.

●○○○○ How many sheepdogs does the farmer need to watch 200 sheep? (2 sheepdogs) To watch 300 sheep? (3 sheepdogs) . . .

●●○○○ How many sheepdogs does the farmer need to watch a flock of 500 sheep? (5 sheepdogs) A flock of 1,000 sheep? (10 sheepdogs) . . .

●●●○○ A sheep farmer has 4 sheepdogs. About how many sheep does this farmer have? (400 sheep) If a farmer has 7 sheepdogs, about how many sheep does the farmer have? (700 sheep) . . .

●●●●○ A farmer has 300 sheep, but only 1 sheepdog. How many more sheepdogs will the farmer need? (2 sheepdogs) If a farmer has 600 sheep and only 2 sheepdogs, how many more sheepdogs will he need? (4 sheepdogs) . . .

Bird Eggs and Their Sizes

Ostrich eggs are about 6 to 8 inches long. Hummingbird eggs are less than $\frac{1}{2}$ inch long.

●○○○○ Which is longer, an ostrich egg or a hummingbird egg? (an ostrich egg)

●●○○○ How much longer than a half-inch long hummingbird egg is a 6-inch long ostrich egg? ($5\frac{1}{2}$ inches longer) How much shorter than an 8-inch long ostrich egg is a half-inch long hummingbird egg? ($7\frac{1}{2}$ inches shorter)

●●●○○ How many half-inch hummingbird eggs placed end to end are as long as an 8-inch long ostrich egg? (16 hummingbird eggs)

●●●●○ How many times longer than a half-inch hummingbird egg is a 6-inch long ostrich egg? (12 times longer) How many times longer than a half-inch hummingbird egg is an 8-inch long ostrich egg? (16 times longer)

Human Water Needs

The average person could survive 6 days without water if the outdoor temperature were around 60°F.

●○○○○ Could an average person survive a week without water if the outdoor temperature were around 60°F? (no)

●●○○○ About how many days could 2 average people survive without water if the outdoor temperature were around 60°F? (still about 6 days)

●●●○○ About how many hours could an average person survive without water if the outdoor temperature were around 60°F? (144 hours)

●●●●○ Could an average person survive more or less than 200 minutes without water if the outdoor temperature were around 60°F? (more) More or less than 1,000 minutes? (more) More or less than 5,000 minutes? (more) More or less than 10,000 minutes? (more)

Short and Tall Women*

The shortest woman on record, Pauline Musters, was about 1 foot 11 inches tall. The tallest woman, Jane Bunford, was about 7 feet 11 inches tall.

●○○○○ To the nearest foot, how tall was Pauline Musters? (2 ft) To the nearest foot, how tall was Jane Bunford? (8 ft)

●●○○○ If Jane and Pauline stood back to back, what would be the vertical distance between the tops of their heads? (6 ft)

●●●○○ The average height for American women is about 5 feet 4 inches. How much taller than the average woman was Jane? (2 ft 7 in.) How much shorter than the average woman was Pauline? (3 ft 5 in.)

* It would be helpful to write these heights on the chalkboard.

Extreme Temperatures

As of 1980, the hottest temperature ever recorded on Earth was about 136°F; this was in Libya. The coldest temperature recorded was about −127°F; this was in Antarctica.

●○○○○ Is −127°F warmer or colder than −130°F? (warmer) Is −127°F warmer or colder than −100°F? (colder)

●●○○○ About how many degrees Fahrenheit is it today? About how many degrees warmer was the hottest day in Libya?

●●●○○ About how many degrees hotter was the hottest day in Libya than the coldest day in Antarctica? (263°F)

Variation Ask similar questions using Celsius temperatures. The hottest temperature was 56°C; the coldest was −88°C.

Hummingbird Wing Flaps

A small hummingbird can beat its wings about 70 times per second.

●○○○○ How many times can you flap your arms in 1 second? (maybe 1 time)

●●○○○ About how many times can a hummingbird beat its wings in 2 seconds? (140 times) In 3 seconds? (210 times) About how many times can a hummingbird beat its wings in half a second? (35 times) . . .

●●●○○ About how many times can a hummingbird beat its wings in 1 minute? (4,200 times) In 2 minutes? (8,400 times) . . .

NUMBER STORIES

Melting Ice Cream

At room temperature, it takes a scoop of ice cream about 15 minutes to melt.

●○○○○ What does “room temperature” mean? (The temperature of an average room; not too hot, not too cold; probably about 70°F or about 20°C)

●●○○○ If the room temperature is 80°F (about 26°C), will it take more or less than 15 minutes for a scoop of ice cream to melt? (less than 15 minutes) If the room temperature is about 60°F (about 15°C), will it take more or less than 15 minutes for a scoop of ice cream to melt? (more than 15 minutes)

●●●○○ About how long would it take 2 scoops of ice cream at room temperature to melt? (Just over 15 minutes; it would not take twice as long just because there are 2 scoops)

Airplane Speed vs. Human Speed

Jets travel an average of 500 miles per hour. People walk an average of 5 miles per hour.

●○○○○ About how many miles could an airplane travel in 3 hours? (1,500 mi) About how far could a person walk in 3 hours? (15 mi)

●●○○○ About how many miles per hour faster do airplanes travel than people who are walking? (495 miles per hour) About how many miles farther could an airplane travel in 3 hours than a person walking? (1,485 mi)

●●●○○ About how many miles could an airplane travel in 8 hours? (4,000 miles) About how many miles could a person walk in 8 hours? (The mathematical answer is 40 miles, but most people could not continue to walk 5 miles per hour for that long.)

●●●●○ About how many times faster does an airplane travel than a person walking? (100 times faster)

Alligator Eggs

A female alligator lays an average of 40 eggs each time she has young.

●○○○○ Do some female alligators lay more than 40 eggs when they have young? (yes) How do you know? (40 eggs is an average; that means some females lay more eggs than the average and some lay less.)

●●○○○ Is the average number of eggs laid by a female alligator more or less than 1 dozen? (more) More or less than 2 dozen? (more) More or less than 3 dozen? (more) More or less than 4 dozen? (less) How many less? (8 less)

●●●○○ What is the average number of eggs laid by 2 female alligators? (80 eggs) By 3 female alligators? (120 eggs) . . .

Candy Sale

Suppose that the grocery store is selling candy bars 3 for $1.00.

●○○○○ How many candy bars could you buy for $2.00? (6 candy bars) How many candy bars could you buy for $3.00? (9 candy bars) . . .

●●○○○ About how much would 1 candy bar cost? (34¢; explain that stores round up to the nearest penny) About how much would 2 candy bars cost? (67¢)

●●●○○ If you had 3 quarters, 4 nickels, and 5 pennies, how many candy bars could you buy? (3 candy bars) If I had 5 dimes, 8 nickels, and 45 pennies, how many candy bars could I buy? (4 candy bars) . . .

Making Pancakes

You need about $\frac{1}{4}$ cup of pancake batter to make one 4-inch pancake.

●○○○○ Will $\frac{1}{2}$ cup of batter make a pancake that is bigger or smaller than 4 inches in diameter? (bigger) Will $\frac{1}{8}$ cup of pancake batter make a pancake that is bigger or smaller than 4 inches in diameter? (smaller)

●●○○○ How many 4-inch pancakes could you make with 1 cup of batter? (4 pancakes) With 2 cups of batter? (8 pancakes) With 3 cups of batter? (12 pancakes)

●●●○○ If you have 1 pint of batter, how many 4-inch pancakes could you make? (8 pancakes) If you have $1\frac{1}{2}$ pints of batter, how many 4-inch pancakes could you make? (12 pancakes) . . .

●●●●○ If each member of your family wanted 4 pancakes for breakfast, how many cups of batter would you need? If each member of this class wanted 2 pancakes for breakfast, how much batter would you need? . . .

A Pint Is a Pound

For water, it is said that "a pint is a pound the world around."

●○○○○ If we had 2 pounds of water, about how many pints would we have? (2 pt, or 1 qt) If we had 3 pounds of water, about how many pints would we have? (3 pt)

●●○○○ About how much would 1 quart of water weigh? (2 lb) About how much would 2 quarts of water weigh? (4 lb) . . .

●●●○○ About how much would a gallon of water weigh? (8 lb) About how much would 3 gallons of water weigh? (24 lb) . . .

●●●●○ There are 2 cups in a pint. About how much would 1 cup of water weigh? ($\frac{1}{2}$ lb) Half a cup of water? ($\frac{1}{4}$ lb)

Bargain Shopping

Which is the bargain?

Pretzels: $0.10 each **or** 3 for a quarter (3 for $0.25)

Gum: a nickel a piece **or** 5 for a quarter (Neither. Both cost the same.)

Pop: a small for $0.50 **or** a large for $0.75 (Insufficient data to answer the question. You don't know the quantities involved in "small" and "large.")

Licorice sticks: $0.15 each **or** 6 for $1.00? ($0.15 each)

Pizza: $1.50 per slice **or** a 6-slice pizza for $8.50? (6-slice pizza)

Long-Distance Telephone Calls

When you made a long-distance telephone call in the past, the first 3 minutes were often the most expensive. For example, if you called New York from Chicago, it would cost you 50¢ per minute for the first 3 minutes. It would cost you 10¢ per minute for each additional minute.

●○○○○ What would a one-minute call to New York have cost? (50¢)

●●○○○ What would a 3-minute telephone call to New York have cost? ($1.50) A 4-minute call? ($1.60) A 5-minute call? ($1.70) . . .

●●●○○ If you were calling New York from a pay phone and you had 4 quarters, how long could you have talked? (2 min) If you had 2 quarters, 10 dimes, and 4 nickels, how long could you have talked? (5 min)

●●●●○ If you had $2.00 to spend on a phone call, what is the longest that you could have talked to New York? (8 min) If you had $3.00 to spend? (18 min) . . .

Popping Corn

Typically, you get approximately 34 cups of popcorn from 1 cup of kernels.

●○○○○ Why does the popcorn take up more space than the kernels?

●●○○○ About how many cups of popcorn will you get if you pop $\frac{1}{2}$ cup of kernels? (17 c) if you pop about 2 cups of kernels? (68 c)

●●●○○ If you wanted about 100 cups of popcorn, how many cups of kernels should you pop? (about 3, or a little less than 3 c)

●●●●○ About how many people can you feed if you pop 1 cup of kernels? (Insufficient data to answer the question. Children may want to estimate anyway.)

Dinosaur Sizes*

The triceratops, a dinosaur with 3 horns, was about 8 to 10 feet tall and about 20 to 25 feet long.

●○○○○ Discuss why scientists would measure the height and length of the triceratops. Would we measure both of these for a person? (no) For a horse? (yes) What are some possible heights and lengths for a triceratops? (8 ft tall and 20 ft long, 8 ft tall and 21 ft long, 8 ft tall and 22 ft long, . . .)

●●○○○ How many horns do two triceratops have? (6 horns) 3 triceratops? (9 horns) . . .

●●●○○ If a triceratops was about 8 feet tall and about 20 feet long, how much longer was it than it was tall? (12 feet) . . .

●●●●○ If a triceratops was 8 feet tall and 24 feet long, how many times longer was it than it was tall? (3 times longer) . . .

Variation Ask questions about the stegosaurus, which was 12–13 feet tall and 18–25 feet long.

* It would be helpful to write the information about the triceratops on the chalkboard.

NUMBER STORIES

Earth's Orbit

The earth travels about 18 miles per second in its orbit (path) around the sun.

●○○○○ Does the earth travel more or less than 18 miles in one minute? (more) Does the earth travel more or less than 18 feet in one second? (more) . . .

●●○○○ About how far does the earth travel in 2 seconds? (36 mi) In 3 seconds? (54 mi)

●●●○○ The earth has moved 80 miles. Have more or less than 5 seconds passed? (less) The earth has moved 100 miles. Have more or less than 3 seconds passed? (more)

Bacteria Growth

If bacteria are given food and a warm, wet place in which to grow, they can double their number about every half hour.

●○○○○ If 100 bacteria live under these conditions, about how many will there be in half an hour? (200 bacteria) In an hour? (400 bacteria) In 2 hours? (1,600 bacteria)

●●○○○ If there are 60 of these tiny creatures, about how many were there half an hour ago? (30 bacteria) 1 hour ago? (15 bacteria)

●●●○○ If there are 10 bacteria, about how long will it be until there are 160 bacteria? (2 hr) If there are 20 bacteria, about how long will it be until there are 160 bacteria? ($1\frac{1}{2}$ hr) . . .

A Dog's Age

One rule of thumb for finding a dog's equivalent human age says to count 7 human years for each year of the dog's life.

●○○○○ If a dog is about 1 year old, what is its "human age" using this rule? (7 years old) If a dog is about 2 years old, what is its "human age"? (14 years old)

●●○○○ If a dog's equivalent human age is 21 years, about how many years has it been alive? (3 years) If its equivalent human age is 49 years, about how many years has the dog been alive? (7 years)

●●●○○ Using "times" or "multiply," restate this rule of thumb. (Multiply the dog's age by 7 to find its equivalent "human age.")

●●●●○ Which do you think usually live longer, dogs or humans? (humans) How can you tell by this rule of thumb? (Each year of the dog's life counts for several years of human life.)

Another Dog's Age

A new rule of thumb for finding the equivalent "human age" of a dog takes into account the fact that a dog can have puppies at 1 year and is full-grown by 2 years. The rule suggests counting the first year of a dog's life as 15 human years, the second as 10, and each year after as 5.

●○○○○ Using the new rule, what is the "human age" of a 1-year-old dog? (15 years old) Of a 4-year-old dog? (35 years old) . . .

●●○○○ Using the new rule, about how many years has a dog been alive if its equivalent human age is 50? (7 yr) If its equivalent human age is 65? (10 yr) . . .

●●●○○ Consider the old rule of thumb which said, "To find a dog's equivalent human age, multiply the dog's actual age by 7." Are the old and new estimates of the dog's equivalent human age closer when the dog is actually 1 year old, 5 years old, or 10 years old? (At the actual age of 1, the estimates are 8 years apart; at the actual ages of 5 and 10, the estimates are 5 years apart.)

Record-Breaking Pizza

The largest pizza ever baked was 100 feet 1 inch in diameter. It was cut into 94,248 slices.

●○○○○ What is a diameter? (The distance from one side of a circle to the other, passing through the center) What do you think would be the diameter of a normal medium-sized pizza? (about 10 in.)

●●○○○ If everyone wanted at least 2 slices, could the largest pizza ever baked have fed 50,000 people? (no) 40,000 people? (yes) If everyone wanted at least 3 slices, could the pizza have fed 30,000 people? (yes)

●●●○○ About how many inches long was the diameter of this pizza? (1,200 in.) About how many inches long was the radius of the pizza? (600 in.)

●●●●○ If you had wanted to put this pizza on a table, how big would the table have had to be? (more than 7,500 sq ft in area)

Human Languages*

About 309 million people in the world speak English. About 322 million people speak Spanish. About 873 million people speak Mandarin Chinese.

●○○○○ Of these 3 languages, which language do the most people speak? (Mandarin) Which language has fewer speakers than the other two? (English)

●●○○○ If we wanted to write the number of people who speak English in the world (309 million), how many zeros would we put after the 309? (6 zeros) How would we write the number of Mandarin speakers? (873 with 6 zeros)

●●●○○ About how many more people speak Mandarin than English? (564 million people) About how many fewer people speak English than Spanish? (13 million people) About how many more people speak Mandarin than Spanish? (551 million people)

* It would be helpful to post the number of people who speak each language on the chalkboard.

NUMBER STORIES

Whale Speeds

The sperm whale travels from 3 to 5 miles per hour if it is left undisturbed. When it is being hunted, the sperm whale can swim at a speed of more than 13 miles per hour.

●○○○○ If a sperm whale is swimming undisturbed, is it likely to swim more than 10 miles in 2 hours? (no)

●●○○○ About how many miles per hour faster does a sperm whale travel when it is being hunted than when it is left undisturbed? (8–10 miles per hour)

●●●○○ If a sperm whale is left undisturbed, about how many miles will it swim in 3 hours? (9–15 mi) If it is being hunted, about how many miles might it travel in the same amount of time? (more than 39 mi)

Koala Birthweights

A newborn koala weighs about $\frac{1}{10}$ of an ounce when it is born. This is about as heavy as one penny.

●○○○○ You have 2 pennies in one hand. How many newborn koalas would you have to put in the other hand to have about the same weight in each hand? (2 koalas) If you added another newborn koala to one hand, do you think that you could feel the difference? (probably not)

●●○○○ About how many newborn koalas would weigh about 1 ounce? (10 koalas) About how many pennies would weigh about 1 ounce? (10 pennies)

●●●○○ About how many newborn koalas would weigh about 2 ounces? (20 koalas) About how many newborn koalas would weigh about 1 pound? (160 koalas)

●●●●○ If a full-grown koala weighs about 20 pounds, about how many times heavier is a full-grown koala than a baby koala? (about 3,200 times heavier) About how many pennies would weigh 20 pounds? (3,200 pennies)

Baseball Speeds

A baseball line drive travels about 100 yards in 4 seconds.

●○○○○ About how many yards does a line drive travel in 2 seconds? (50 yd) In 1 second? (25 yd) . . .

●●○○○ About how many feet does a line drive travel in 4 seconds? (300 ft) About how many inches does a line drive travel in 4 seconds? (3,600 in.) . . .

Age: People vs. Mice

Many mice live to be 4 years old. Many people live to be 80 years old.

●○○○○ Are there many mice as old as you? (no) Are there many people as old as you? (yes) How many birthdays do most mice have? (4 birthdays) How many birthdays have you had?

●●○○○ About how many years longer than a mouse do many people live? (76 yr) About how many years longer will you live, if you live to be 80 years old?

●●●○○ How long is half of a 4-year-old mouse's life? (2 yr) About how long is half of an 80-year-old human's life? (40 yr) About how long is half of your life so far?

●●●●○ About how many times as long as a mouse do many people live? (20 times as long)

People Heat

It is said that 10 people will raise the temperature of a medium-sized room about 1°F in one hour.

●○○○○ Would 10 people raise the temperature of a large room more or less than 1°F in an hour? (less) Would 10 people raise the temperature of a small room more or less than 1°F in an hour? (more)

●●○○○ If there are 10 people in a medium-sized room and the temperature has risen 3°F, about how many hours have they probably been in the room? (about 3 hr) If the temperature had risen 6°F? (about 6 hr)

●●●○○ If there were more than 10 people in a medium-sized room, would you expect the temperature to increase more or less than 1°F per hour? (more) If there were 20 people in a medium-sized room, about how many degrees Fahrenheit per hour would you expect the temperature to rise? (2°F per hour)

Water in an Elephant Trunk

The trunk of an average elephant can hold about $1\frac{1}{2}$ gallons of water.

●○○○○ About how many milk cartons of water can the average elephant hold in its trunk? ($1\frac{1}{2}$ of the big plastic gallon containers; 3 of the medium-sized cardboard half-gallon containers)

●●○○○ About how much water can the trunks of 2 average elephants hold? (3 gal) Of 3 average elephants? ($4\frac{1}{2}$ gal)

●●●○○ How many average elephants would be needed to hold about 3 gallons of water? (2 elephants) To hold about 6 gallons of water? (4 elephants) . . .

●●●●○ About how many quarts of water can an average elephant hold in its trunk? (6 qt) About how many pints of water can an average elephant hold in its trunk? (12 pt)

List of Activities by Page

Basic Routines **2**
Numbers Before and After 3
Numbers Between 4
Counts and Skip Counts 5
Ordinal Numbers. 6
Count by 10s and 100s 7
What Do I Do? 8
Complements of 10s 9
Arithmetic Facts 10
Name Collections (Equivalents) 11
More Name Collections (Equivalents) 12
Multistep Problems 13
How Many 10s, 100s, 1,000s? 14
What's My Rule? 15
Number Stories 16
Shapes Around Us. 17
Geometry I Spy 18
Comparing Measurement Units. 19
Recent Dates 20
Unit Conversions. 21

How Many Cents? 22
Place Value 23
Ten More, Ten Less. 24

Minute Math Topics. 25
Counting 27
Numbers Before and After 27
Whole Numbers Between 28
Missing Numbers 29
How Many? 30
Continue the Sequence 31
Repeated Digits 32
Numbers with *n* Digits 33
Easier Numbers. 34
Money and Measure Counts 35
Creating Numbers. 36
Place Value in Metric Measures. 37
Guess My Number. 38

Operations. 39
Missing Parts in Sums and Differences. 39
Parts in a Whole 40

Fact Families . . . 41
Addition and Subtraction Properties of 10s . . . 42
Multiplication Properties of 10 . . . 43
How Many Tenths, Hundredths, Thousands? . . . 44
Digit Arithmetic . . . 45
Secret Numbers . . . 46
Siblings . . . 47
Double, Triple, Quadruple . . . 48
Parts . . . 49
Estimating Differences . . . 50
Tenths, or 10% . . . 51

Geometry . . . 53
Imagining Shapes . . . 53
Uni-, Bi-, Tri- . . . 54
Shapes of Signs . . . 55
Making Shapes and Angles . . . 56
Identifying Line Segments . . . 57
Describing Shapes . . . 58
Walking Shapes . . . 59
Class Shapes and Lines . . . 60

Measurement 61
Measuring Tools 61
Informal Measuring Tools 62
How Far to Special Dates 63
Making a Dollar 64
Money Exchanges 65
More Money Exchanges 66
Coin Equivalents 67
Money and Time 68
How Many? 69
Again, How Many? 70
Telling Time 71
Time Zones 72
Understanding Time 73
Equivalent Measures 74
Measurement Fractions 75

Number Stories 77
Baby Penguin Meals 79
Sleep Needs 80
Thunder 81
Rug Measures 82

Old Milk 83
Dreams 84
Airplane Pilots 85
The Length of a Dollar Bill 86
Koala Bears 87
Classroom Counts 88
Hot Dogs and Buns 89
Making Omelets 90
Outgrowing Shoes 91
Game Time 92
Toilet Flushes 93
Packs of Gum 94
Making Apple Juice 95
Left-Handed People 96
Outdoor Temperatures 97
Making Orange Juice 98
Cooking-Oil Consumption 99
Making Potato Salad 100
A Snail's Pace 101
Riddles 102
The Oldest Living Animal 103
Lifespan of a Dollar 104

Baby Weights . . . 105
Planting Flower Bulbs . . . 106
New Books . . . 107
Patents for Inventions . . . 108
Elephant Meals . . . 109
Elephant Sleep . . . 110
Hibernation . . . 111
Making Cookies . . . 112
Types of Bears . . . 113
The Three-Toed Sloth . . . 114
Pumping Blood . . . 115
Basketball Player Heights . . . 116
Body Muscle . . . 117
A Long Monopoly® Game . . . 118
Paper in Trash Heaps . . . 119
Breathing . . . 120
Cricket Volumes . . . 121
Tall Buildings . . . 122
Walking Rates . . . 123
Penguin vs. Human Swimmers . . . 124
Length of a Shrew . . . 125
Animal Litters . . . 126

Washing and Drying Clothes 127
Growing Eyelashes 128
Animal Weights 129
Body Water Weight 130
Milk Consumption 131
Herding Sheep 132
Bird Eggs and Their Sizes 133
Human Water Needs 134
Short and Tall Women 135
Extreme Temperatures 136
Hummingbird Wing Flaps 137
Melting Ice Cream 138
Airplane Speed vs. Human Speed 139
Alligator Eggs 140
Candy Sale 141
Making Pancakes 142
A Pint Is a Pound 143
Bargain Shopping 144
Long-Distance Telephone Calls 145
Popping Corn 146
Dinosaur Sizes 147
Earth's Orbit 148

Bacteria Growth 149
A Dog's Age 150
Another Dog's Age 151
Record-Breaking Pizza 152
Human Languages 153
Whale Speeds. 154
Koala Birthweights 155
Baseball Speeds. 156
Age: People vs. Mice 157
People Heat 158
Water in an Elephant Trunk 159

Key to Sources

Minute Math+ page	Source
81	Sandow, p. 18
84	Smith and Moore, p. 84
91	Parker 1987, p. 73
93	Smith and Moore, p. 74
96	Smith and Moore, p. 28
98	Parker 1983, p. 9
104	Smith and Moore, p. 127
105	Parker 1987, p. 23
106	Parker 1983, p. 20
107	Parker 1984, p. 15
109	Smith and Moore, p. 99
110	Balfanz, p. 67
111	McFarlan, p. 36
111	Stephen, p. 152
113	Balfanz, p. 45

Minute Math$^+$ page	Source
114	McFarlan, p. 35
115	Parker 1987, p. 41
116	McFarlan, p. 344
117	Parker 1987, p. 9
118	Monopoly® homepage
119	Parker 1987, p. 6
120	Parker 1987, p. 48
121	Parker 1987, p. 89
122	John Hancock Center Observatory
123	Smith and Moore, p. 34
125	Diagram Group, p. 71
126	Burton, p. 84
128	Smith and Moore, p. 10
129	Diagram Group, p. 119
133	Diagram Group, p. 69
134	Smith and Moore, p. 17
135	Diagram Group, p. 73
136	Diagram Group, p. 151

Minute Math⁺ page	Source
137	Sandow, p. 12
138	Sandow, p. 52
140	Smith and Moore, p. 107
146	Parker 1987, p. 15
147	Diagram Group, p. 65
150	Parker 1987, p. 282
151	Parker 1987, p. 282
152	McFarlan, p. 259
153	McGeveran, p. 731
154	Stephen, p. 306
156	Smith and Moore, p. 199
158	Parker 1983, p. 113
159	Smith and Moore, p. 101

Bibliography

Balfanz, Robert. *Do Elephants Eat Too Much?* Evanston, IL: Everyday Learning Corporation, 1992.

Boehm, David A. (Ed.) *Guinness Sports Record Book 1990–91.* New York: Sterling Publications, 1990.

Burton, Maurice (Ed.) *The World Encyclopedia of Animals.* New York: Funk & Wagnalls, 1972.

Diagram Group. *Comparisons.* New York: St. Martin's, 1980.

Hasbro, Inc. Monopoly® homepage. (online). http://www.hasbro.com/pl/page.funfacts/dn/default.cfm (accessed December 2005).

John Hancock Center Observatory. Information provided to visitors. Chicago.

McFarlan, David A. (Ed.) *Guinness Book of World Records 1990.* New York: Sterling Publications, 1989.

McGeveran, William A., Jr. (Ed.) *The World Almanac and Book of Facts 2006.* New York: World Almanac Books, 2006.

Parker, Tom. *Rules of Thumb.* Boston: Houghton Mifflin, 1983.

_____. *In One Day.* Boston: Houghton Mifflin, 1984.

_____. *Rules of Thumb 2.* Boston: Houghton Mifflin, 1987.

Sandow, Stuart A. (Ed.) *Durations: The Encyclopedia of How Long Things Take.* New York: Avon, 1977.

Smith, Richard, and Moore, Linda (Eds.) *The Average Book.* New York: The Rutledge Press, 1981.

Stephen, David (Ed.) *Encyclopedia of Animals.* New York: St. Martin's, 1968.